Also by David Cook

When Your Child Struggles: The Myths of 20/20 Vision

Visual Fitness: 7 Minutes to Better Eyesight and Beyond

The Anatomy of Blindness: A Novel

BIOMYTHOLOGY

The Skeptic's Guide to Charles Darwin
and the Science of Persuasion

David Cook

authorHOUSE®

AuthorHouse™
1663 Liberty Drive
Bloomington, IN 47403
www.authorhouse.com
Phone: 1 (800) 839-8640

Published by AuthorHouse 04/27/2016

ISBN: 978-1-5246-0183-6 (sc)
ISBN: 978-1-5246-0181-2 (hc)
ISBN: 978-1-5246-0182-9 (e)

Library of Congress Control Number: 2016905559

Print information available on the last page.

This book is printed on acid-free paper.

Illustration Brian Farmer
Cover Photography: Copyright 2013
Sawyer Photography
Used with permission.

To A. R. Braunmuller:

Thanks for the introduction to eloquence.

Contents

[T]here must be a separation of state and science just as there is a separation between state and religious institutions, and science should be taught as one view among many and not as the one and only road to truth and reality.

—Paul Feyerabend, *Against Method*

So humanities, farewell. You do not survive Darwinian cost-benefit analysis.

—Marilynne Robinson, *The Givenness of Things*

Introduction

Are you skeptical when the miracles of physical science are cited to alter your worldview rather than the world you view? Do you believe that, without the humanities along for the ride, science can no more capture the mystery of what it means to be human than a lone child can discover the joy of a seesaw—the soaring, the caress of sky, the transcendence? If so, this book is for you.

Full Disclosure

In this training manual for skeptics, *Biomythology* is the myth that biology can replace the humanities in capturing what it means to be human. The myth is told using, not Greek, but what philosopher of science Paul Feyerabend has called "brainwashing by argument,"[1] a special form of rhetoric that we will learn to recognize as the *science of persuasion*.

Why center our skepticism on this particular myth? Why doubt Darwin, not the Greeks? Why pit science against the humanities, not just religion? Why detail the rhetorical devices of scientific plain prose used to change minds, not matter? Since one of my contentions throughout these chapters will be that all literature, scientific or otherwise, reflects the biases and background of the author, it is only fair that I reveal my own biases and background.

David Cook

As you know, in philosophy the logic of an argument stands or falls on its own merits and not on the length of the philosopher's rap sheet. Who cares if Sir Francis Bacon was accused of twenty-some counts of bribery, sent to the Tower, fined £40,000, or stripped of his offices?[2] In philosophy, calling attention to such idiosyncrasies is dismissed as an *ad hominem* or "to the man" fallacy (women apparently being unworthy of idiosyncrasies or incapable of fallacies). Science is a different beast, often relying on observations. Where observations are concerned, the integrity of the observer becomes critical. In science, *ad hominem* arguments are encouraged; full disclosure of any potential conflict with the interests of orthodoxy or peer review is now demanded whenever I lecture or write.

With my own conflicted interests in mind, here is my disclosure: I began attending UCLA in 1970. Majoring in English on the aesthetic North campus and minoring in pre-optometry on the sterile South campus, I studied literature because I loved words, science because I loved eating. Exposed to everything from Milton to math and Chaucer to chemistry, I graduated to attend optometry school.

Professional school began slowly. Studying the psychology that would underlie our testing procedures, we reviewed Ivan Pavlov, who taught dogs to slobber to the tune of bells and whistles. In optics, we learned that, because light traveling from twenty feet behaves pretty much like light traveling from the stars, *optical infinity* is twenty feet. I began to have misgivings: not only had I traded the doggerel of Lewis Carroll for the dog drool[3] of Ivan Pavlov, but also I was doomed to enter a profession that was nearsighted enough to confuse twenty feet with infinity—or so I fretted.

Not until my third year did optometry come alive as I attended the classes of Professor William Ludlam. He introduced us to vision therapy.[4] Bill, a popular speaker to optometric audiences throughout much of the English-speaking world, lectured about more than eye conditions; he lectured about the patients he had helped, individuals whose lives he had changed. He insisted that we were on "the side of the angels." Inspired, I traded the enrichment of literature for enriching the lives of patients. I began volunteering ten hours per week in the vision therapy clinic. As Bill promised, I changed lives.

Having earned my Doctor of Optometry degree, I completed a vision therapy residency at the State University of New York and entered private practice to specialize in my new passion. Continuing my lifelong education, I compared the theories I had been taught in school to my own clinical observations. As our philosopher friend Feyerabend wrote, "[T]here is not a single interesting theory that agrees with all the known facts in its domain."[5] Optometry proved no exception.

I refined procedures, observed what worked for patients, and learned how logic is often used to stretch explanations at odds with results. More interestingly, I learned that, just as seeing is not chiseled in neural stone on the back of the eye, there were different explanations from different perspectives for the same observations. Each explanation had its own strengths and weaknesses—what worked for patients was more important than any explanation. The truth rested in the actions, not stories. Breakthroughs often rested in the art of selecting from the stories and actions best suited to the patient at hand. *Hows* became more important than *whys*.

I became skeptical of those promoting the truth of a single perspective in understanding visual consciousness. I began to wonder if the truth of seeing didn't emerge from, but exceed, the sum of the perspectives. I seriously began to doubt if vision, much less human behavior, could be reduced to universal theories of stimulus, response, and statistics.

Along the way, I wrote articles on vision therapy. They were accepted in the peer-reviewed clinical science journals of optometry.[6] I reviewed articles for the *Journal of the American Optometric Association*, wrote exam questions on vision perception for the National Board of Examiners in Optometry, and spoke at international conferences to doctors from across the world on the challenges of visual consciousness in the nonsurgical treatment of strabismus (crossed eyes). Did any of this qualify me as a scientist? No. According to organized medicine, it qualified me as a quack.

In a sense, the charges were true. Contemporary historian of science Frederick Gregory writes, "A quack in the German states of the eighteenth century was someone who poached on the territory of another and in so doing upset the proper economic order."[7] The definition has not really changed. Like *scientific*, *quack* is still evoked primarily for the protection of economic turf. In the minds of those calling me a quack then, optometrists have certainly upset the proper economic order.

There are two eye-care professions: ophthalmology, a medical specialty, and optometry, an independent profession like dentistry. For over a century, these two professions have fought for prestige, power, and patients. Their skirmishes never cease. Organized optometry has used grass-roots politics to increase their scope of practice to include much of ophthalmology's former exclusive eye-disease turf. During the same period, ophthalmology has authored position statements designed to attack optometry's principal contribution to vision care: vision therapy to enhance human performance.

In 1981, for instance, ophthalmology inspired the three-decade run of the policy statement, "Learning Disabilities,[8] Dyslexia[9] and Vision."[10] The original version of the statement weighed in at a page and a half and offered fifteen references to support the conclusion, "No known scientific evidence supports the claims for improving the academic abilities of dyslexic or learning disabled children ... with treatment based on visual training." By 2011, the reissued statement had burgeoned to thirty-nine pages and 279 references, dropping facts and names with a scientific abandon.

Medicine's conclusion, found in the abstract at the beginning of the article, is the only part that a busy pediatrician is likely to read:

> Currently there is inadequate scientific evidence to support the view that subtle eye or visual problems cause or increase the severity of learning disabilities ... Scientific evidence does not support the claims that visual training, muscle exercises, ocular pursuit-and-tracking exercises ... [etc.] are effective direct or indirect treatments for learning disabilities. There is no valid evidence that children who participate in vision therapy are more responsive to educational instruction than children who do not participate.[11]

To emphasize its validity, the abstract uses the word *scientific* twice and *evidence* three times. Curiously, buried in the middle of a paragraph, in the middle of a page, in the middle of the body of the statement, we now find the caveat:

> Symptomatic convergence insufficiency[12] can cause discomfort, eyestrain, blurry vision, ... [double vision,] and headache, which can contribute to limited fluency by interfering with the child's ability to concentrate on print for a prolonged period of time. Symptomatic convergence insufficiency is a treatable condition.[13]

I'll leave it up to you to decide if the relief of eyestrain, double vision, and reduced concentration might directly or indirectly reduce difficulties with learning. The real question is why the caveat was included at all.

As it turns out, convergence insufficiency had to be addressed because of a 2008 gold standard, placebo-controlled[14] study financed and supervised by the National Eye Institute, the government's lead agency for vision research. For those who believe in evidence-based medicine, the study provided strong support for in-office vision therapy being the best treatment for symptomatic convergence insufficiency.

Oddly, the 2011 medical policy statement did not recommend in-office vision therapy; the statement instead referenced inferior studies without placebo controls to question the need for optometric in-office vision therapy. In clinical science, evidence is seldom heavy enough to outweigh prejudices and pocket books. Criticisms against vision therapy had never been about the adequacy of studies. The criticisms had simply confiscated the words *scientific* and *quack* to protect economic turf.

Scientific Skepticism

Thirty years of seeing the adjective *scientific* flaunted to discredit my personal observations and to condemn children to headaches and double vision, simply because they failed to fit theory, turf, and pocketbooks, only reinforced my skepticism about the adjective. In the culture of sales, *scientific* has come to modify everything from literature to laxative commercials. Cosmetologists borrow *scientific* to sell moisturizing cream, physicians borrow *scientific* to sell botulism injections, Darwinians borrow *scientific* to sell materialism, and Creationists borrow *scientific* to sell God.

Those who shout the term most loudly are the most suspect. The further the skeptic strays from physics and chemistry, the more the word appears. It seems that only when a product sells itself can *scientific* be safely dropped. As we will see, the telegraph, telephone, and electric light all appeared in the records of the patient office before they were mentioned in the *scientific* literature. Today, no one wastes time selling the public on the idea that smartphones are *scientific*. When real science is involved, no one has to sell it: a century and a half of continuous promotional efforts are unnecessary.

Which brings us to Charles Darwin.

Darwinism

If we seldom are reminded that it is the consensus of the scientific community that Maxwell's or Schrodinger's or Einstein's equations are true, then why must we be repeatedly be told that all reputable biologists agree that evolution is true? Based on thirty years of dealing with the position statements of scientific academies, the skeptic in me suspects that when the word *scientific* is wed to the word *consensus,* our only certainty is that we are in the middle of a sales pitch. "It is the consensus of the scientific community that" means, quite simply, "Buy now!" To the mind of the skeptic, scientific consensus is little more than a euphemism for political science.

With that suspicion in mind, I began to explore the history and philosophy of science—often from my perspective as a reader of literature and my experience dealing with science as a clinician. Rather than search for scientific truth, I searched for how scientific truth is sold. I found that there is probably no better example of science borrowed for the purpose of sales than Darwinism. While no one wastes time writing books to assure us that Newton or Einstein are true, between 2008 and 2009 there were three books written for popular audiences,[15] all trying to redefine the borders of skepticism, all forbidding the skeptic to doubt that Darwinism is true.

In this book, rather than borrowing Feyerabend's "brainwashing by argument," I capture the process with the less vividly descriptive rhetoric of objectivity or, if you will, *science of persuasion* appearing in our title. I make no attempt to disprove the stories of Darwin and his disciples. I merely question their rhetoric. Such words as *prove* and *disprove* suggest certainty; we can hardly demand skeptics to be certain. Instead, reading Darwin and his disciples, I catalogued, often in my own words, the rhetorical devices used when borrowing the prestige of science to sell ideas to the culture. To name a few: scientific skepticism, plain prose, equivocation, provisional truth, myth expanders, musical definitions, caveats, the silence of jets, fallacy to a higher power, logic, fact-dropping, name-dropping, footnotes, Cat-in-the-Hat empiricism, magical infinite thinking, rhetorical pictures, rhetorical math, insignificant figures, rock-solid observations, and the universal

David Cook

solvent of rhetorical reason washing away all observations at odds with passion.

Luckily, if skepticism is true, then the doubter is always free to doubt skepticism itself. To doubt or not to doubt remains the question. This book will ultimately argue that it's time to doubt a story, scientific or otherwise, when that story is used to justify actions that, in the philosopher's words, make "the life of man, solitary, poor, nasty, brutish, and short."[16] For instance, the 169,198,000[17] human beings murdered in the twentieth century in the name of government ideologies (many with *scientific*[18] pretensions underlying their approaches to "progress") could have benefited greatly from a public schooled in skepticism.

Lest, in the future, the stories of science can again be used to justify actions we cannot take back, the chapters that follow will examine, under the glass of skepticism, stories told about reason, evidence, experiment, and observation, making it easy for the skeptic to doubt should skepticism, without a doubt, again become necessary.

It was the objectivity of Sir Isaac Newton (1643–1727) that inspired the miracles and hope that earned real science the respect it deserved. Newton allowed, "If I have seen farther it is by standing on ye shoulders of giants."[19] If, like Newton, we are to see farther, must we be willing to tread on the tales of biomythology? The skeptic should be trained to ask if using Darwin's bestselling words to explain away what makes us human is as risky as using the space between atomic particles to explain away the solidity of an approaching brick. Join us as we review for the skeptic how the words of universal explanations too often dissolve into myth when transported from one universe to another—myth and pain.

Chapter 1

Skepticism without a Doubt

The most perfect philosophy of the natural kind only
staves off our ignorance a little longer.[20]
—David Hume (1711-1776)

Like big government, big science grows stronger every
day, passing laws to constrain every facet of nature. No fact, no
observation, is above the law. As a result, the prisons of intellectual
history are overflowing with scientific facts gone bad. Waves of light
break into quantum sand. Time and space genuflect to viewpoint.
Pluto lost the vote.

If our elementary school science texts already contain errors,
where exactly do we draw the line between theory and myth? Will
today's facts withstand the ravages of discovery and consensus to
outperform those former facts that now sound like fairy tales, such
as "Once upon a time, there were nine planets"? If you find your
skepticism is increasing, read on.

All Things Are Possible to Doubt

Skepticism is doubt. Doubt can be dangerous. Ask anyone
in sales. Whether your product is faith, rationality, or genocide,
skepticism kindles hesitation, ignites freedom of thought, and
dampers the impulse to buy. The rallying cry of those who are intent

on slaughtering or demeaning their fellows is seldom "I'm just not certain!" It is small wonder that skepticism has gotten a bad name.

Make no mistake: skepticism is not cynicism. The skeptic has doubts about the folly of human passions; the cynic has no such doubts. What's more, while cynics lack faith, true skeptics are not ashamed to admit that—ultimately knowing nothing—they must act on faith in something: love, logic, explanations voted most likely to succeed, beauty, dreams, inspiration, justification, habit, reason, observation, consensus, pragmatism, the human spirit, the Holy Spirit, the sanctity of human life, etc.

Faith makes the skeptic's world go round. The skeptic suspects it is vanity to confuse a web of belief with a network of knowledge. Most importantly, without having to fret about knowledge, skepticism can be fun! In this book, we are going to have our fun and poke it too—at those borrowing the name and tools of natural science to sell ideas to our culture rather than to increase our control over nature.

Divided Skepticism

We may divide skeptics into two camps: turf skeptics and real skeptics.[21] While the mutual doubt between Republicans and Democrats epitomizes turf skepticism, real skeptics doubt politics and politicians. While real skeptics doubt knowledge in general, turf skeptics can be counted on to doubt the knowledge of their competitors.

2

Real skeptics believe that knowledge requires certainty. They claim you cannot really *know* if the sun is shining unless you are *certain* that the sun is shining. How certain? Real skeptics claim that certainty is absolute. It is like the plague: either you have it or you don't. If you are less certain of something than the nose on your face, you are not certain about it at all. Real skeptics not only think outside the box, but some doubt the box exists outside the stories of men and women. For real skeptics, certain knowledge hangs on the thread of certain faith in the death of inconvenient future discoveries.

Real skeptics are rare. Turf skeptics abound. Orthopedic surgeons are skeptical about chiropractors; Protestants, about Catholics; atheists, about theists. All qualify as turf skeptics. Wherever there is turf, a turf skeptic is never far away.

Take professional skeptic Michael Shermer, who writes the "Skeptic" column for *Scientific American*. He is dismayed to cite a 2009 Harris Poll of adult Americans revealing that 82 percent believed in God and only 45 percent believed in Darwin's Theory of Evolution.[22] Shermer may be the king of skeptics, but he's hardly as skeptical as skeptical can be. He is a scientific skeptic, a turf skeptic defining skepticism as "the application of the scientific method to test all claims."[23] He is obviously no skeptic about the superiority of the scientific method despite there being no prospective, placebo-controlled studies to prove the method's superiority over the humanities in discovering what it means to be human. Shermer's

3

brand of skepticism requires us to reach for our pocket calculators each time we hear the words "I love you!" How else can we know if the claim is statistically significant? Such is the peril of abandoning our doubts to the provincial whims of scientific turf skepticism.

Leveling the Playing Field of Skepticism

The golden age of skepticism has passed as surely as the Wild West. Consider Scottish philosopher David Hume, quoted above. Now there was a real skeptic, a radical skeptic, and not some namby-pamby modern-day skeptic short on doubt that the scientific method can do more than temporarily blind us to our ignorance. Hume might have joined us in asking how we can rationally believe that Mother Nature is flipping a two-headed coin—heads yesterday, heads today, heads tomorrow. How can we be rationally certain that our run of luck is not about to end? Real skeptics like Hume caution that we predict the future from the habits of the past at our rational peril. Forget probability. Long shots happen. No matter what the odds, a horse can't win by a nose when the nose is still shy of the finish line; the truth of the horse race is in the photo finish, which transcends both the race and the odds.

What was the probability that two skyscrapers towering over New York for more than ten thousand days would no longer be standing at the end of September 11? Despite the rightful shock and sorrow still attached to this tragedy, Hume might have agreed with the analogy that scientific evidence and certain reason are about as rock solid as a World Trade Center: waiting to succumb to the terror of future discovery.

Hume left us with plenty of room for doubt about the providence of both the natural and supernatural. Today, Hume's brand of skepticism is vanishing. Stories of the supernatural monopolize our doubt, leaving little left for stories of the natural. As a result, skepticism is losing its balance. Doubt threatens to topple into the abyss of certainty. In life, balance is everything. Novelist and scientist C.P. Snow (1905-1980), in his famous 1959 public Cambridge lecture "The Two Cultures," bemoaned the lack of balance in British culture. Supposed intellectuals, Snow noted, could quote Shakespeare but not the most basic laws of physics. The skeptic is rightly concerned about the similar growing asymmetry in skepticism. Therefore, to regain equilibrium in our doubt, welcome to Skepticism 102: *Biomythology: The Skeptic's Guide to Charles Darwin and the Science of Persuasion.*

In Skepticism 101, *The God Delusion*, renowned Darwinian apologist Richard Dawkins taught the skeptic's perspective on the stories of the supernatural. One-time Charles Simonyi Professor for Public Understanding at Oxford University, Dawkins is a rhetorical genius when it comes to selling science to the public. Similarly, when it comes to the light of skepticism, Dawkins is the brightest wavelength on the spectrum. *Delusion* provided a litany of arguments to popularize skepticism about religion. The polemic was scathing and exhaustive, if not exhausting. We will, therefore, ignore skepticism about the stories of the supernatural and explore, instead, skepticism about stories of the natural. In other words, we will examine skepticism without a doubt—the doubt about transcendence. All other doubts, and doubters, we will happily entertain.

In ages past, skepticism about religion caused ears and hearts—among other organs—to burn. Today, as science and religion vie for our belief, doubts about science cause their own share of heartburn, not to mention conflagrations in the meeting rooms of peer-review and tenure committees. Apologizing in advance for any inadvertent fires, *the goal of this book is to level the playing field of the game of skepticism, doing for Darwin and science what Richard Dawkins has already attempted for God and religion.*

Along the way, we will mention no intelligent designs except for the designs of those using the name of science to sell something. Similarly, we will not refute Darwinism—only examine Darwinian rhetoric. The words playing upon these pages form an instruction manual for recognizing the many devices of the rhetoric of objectivity concealed within scientific plain prose. We will teach the skeptic to pick scientific explanations apart at their artificial joints and to doubt science whenever it is taken out of the context of its own discipline, whether to replace the humanities or to allow cruelty to have its reasons.

Why Doubt?

All doubts are equally dangerous, but some doubts are more equal than others. It depends on what you are selling: eternity or oblivion. Many protested when Dawkins penned words that could be used to cast doubt on religion. Many will protest as I pen words that may be used to cast doubt on science. The skeptic will learn, however, that it is not the words that are dangerous. People, not guns, kill people. People, not words, tell lies.

Since the time of Aristotle, it has been known that the words of rhetoric can be used for good or ill. In the great philosopher's words about words, "A man can confer the greatest of benefits by a right use of these, and inflict the greatest of injuries by using them wrongly."[24] The skeptic suspects that whether words result in benefit or injury rests with actions, not the words themselves. If people could use the reform of Christianity to turn "Love one another!" into a hundred years of Christians killing Christians over the correct definition of *love*, then what idea can't be borrowed to justify passion?

The "this idea could be used for evil" argument could be used against most any scientific, theological, or philosophical idea ever imagined, skepticism included. The Golden Rule, in the hands of a masochist with a death wish, can be used to excuse torture and serial killing. The most exquisite Grecian urn ever spun from a potter's wheel can still, if swung with enough momentum, end life. So it is with the most beautiful prose.

As we will see, the universal solvent of rhetorical reason can be used to wash away all actions at odds with passion. Reasons pass; the consequences of actions persist. When it comes to physical cruelty or killing, the skeptic knows enough to doubt whatever words are used to justify, condone or glorify the act. In Shakespeare's tragedy, Othello killed his sea of kisses, Desdemona, not because of Iago's villainous words, but because the great general failed to doubt the words rather than the act of killing. Othello would have flunked Skepticism 102.

Badly.

The most radical of skeptics, the solipsists, doubt actions and believe only in their own private stories, their silent words and images. In this book, we will counsel the skeptic to doubt stories, not actions. If you doubt a world beyond your personal stories, please respect the rest of us, even if logic tells you we do not exist. Flush the toilet.

Solipsist or not, the real scientist may safely doubt that the words of this book will encourage people to throw out their Blu-ray discs, electric lights, or antibiotics. The public, like Francis Bacon before them, are more concerned with the fruits than the explanations of science. If Darwinism could predict what natural selection will do next, we'd all accept it as a legitimate discipline of *science*. But it can't, so we won't.

The skeptic may suspect that Darwinian efforts to export their explanation from a tool of science to a demand for cultural belief may well have sacrificed the credibility of all science. Largely thanks to Darwinism, the skeptic may imagine, scientific consensus has become confused with "The Boy Who Cried Wolf."

Must we alter our faith and lifestyle because of a boy who keeps changing his stories? Don't blame my polemic if the boy is no longer believed. Don't blame me if our ozone is being strangled by the paw prints of a carbon wolf or the methane gas of a flatulent cow. Personally, I would suggest the skeptic look at the stories themselves, not just the questionable history of the storyteller.

Fundamental Incomprehensibilities

Dangers aside, the skeptic will delight to find that nature provides abundant opportunity for the study of doubt. One form of such doubt is called *agnosticism*, from the Greek *gnōsis* for "knowledge" and the *a* for "without." The agnostic is thus "without knowledge." In the now famous words of "Darwin's bulldog," the man who coined the term *agnosticism*, Thomas Huxley (1825-1895), "In the ultimate analysis everything is incomprehensible, and the whole object of science is simply to reduce the fundamental incomprehensibilities to the smallest possible number."[25]

Defining Science

If the purpose of science is, as Huxley suggests, concentrated "fundamental incomprehensibilities," what in fact is science? As the premature death of countless facts has alerted the skeptic to surmise, it takes more imagination to invent a lasting fact than a lasting aphorism. The curved space of non-Euclidean geometry has long since seduced us from the arms of Euclid's "shortest distance between two points," while Socrates' previously cited "All I know is that I know nothing" shows no signs of being a casualty of progress. Indeed, Socrates' wisdom has become the real skeptic's universal rallying cry. Placing aesthetics before rationality, we will therefore skip the facts of science, for a time, and turn to the aphorisms.

To this skeptic, "Science is the art of arranging observations to fit theory." Happily, better wits than mine have assaulted the topic. English author Samuel Butler (1835-1902) proposed that science "is only an expression for our ignorance of our own ignorance."[26] Nobel Laureate Richard Feynman (1918-1988) wrote, "Science is the belief in the ignorance of experts."[27] Philosopher of science Karl Popper (1902-1994) suggested, "Science may be described as the art of systematic over-simplification."[28] Best of all, Nobel Laureate Ernest Rutherford (1871-1937) captured science with the eloquence and parsimony previously reserved for a good equation: "All science is either physics or stamp collecting."[29]

Rutherford's contributions to physics were a gift to the world. Building on Rutherford's philanthropy and philately, we arrive at a new dichotomy—*real science* versus *biomythology*—and a new set of aphorisms. Real science applies the scientific method to control the physical universe; biomythology borrows the method to control our minds. When it comes to verifying electricity and chemistry,

10

the flick of a plastic light switch is all that is required. The light of biomythology has no switch, only *scientific* consensus.

The good skeptic will quickly note that, as I use it, the term *biomythology* is concentrated fallacy, able to poison any well with a single drop. It plays to our emotions; it is a straw-man term discrediting the arguments of all imagined opponents. It is a hasty generalization, using the example of a few *scientific* proselytizers to discredit the many devoted scientists of consciousness, who have devoted a lifetime in the pursuit of truth. It is a plea for our own illogic, based on the illogic of others. It is a false alternative, forcing readers to choose between reason and faith in defining what makes us human—as if there were anything reasonable or faithful about us. It is an appeal to the majority because "everyone knows" what it means to be human, and just about everyone wishes we wouldn't act that way. It's the opportunity to provide the skeptic with a really good laugh as the changing fashions of the emperor's new clothes fail to cover the warts of the emperor's old theories.

Sure, we could have used less incendiary terms: *natural science* versus *natural philosophy* and *natural persuasion*. But why spare the emotions of those who take pride in having none? Such an attempt would be an insult to any rationalist worthy of the name. Continuing to use the terms *biomythology* versus *real science*, the skeptic need only remember which science insists we see the light and which science provides the light so we can see.

History

For some, history is what happened. For many, history is a celebration of the events forming our culture. For the skeptic, history is a story engineered to reconcile the nightmares of yesterday with our dreams for tomorrow. A review of history teaches the skeptic that concocted or reconstructed terms like *biomythology* are nothing new. Consider an introduction to the father of the Italian renaissance, Francesco Petrarch (1304-1374):

> What is the use—I beseech you—of knowing the nature of quadrupeds, fowls, fishes, and serpents and not knowing or even neglecting man's nature, the purpose for which we are born, and whence and whereto we travel?[30]

Petrarch, in addition to elevating humanity to its correct place above biology, invented the term *Dark Ages* so he could salvage us from the supposed ignorance of medieval scholars. He searched Greek and Roman classics for what distinguishes us from the fowls and fishes to make us human. He launched humanism and the humanities.

Five centuries later, Charles Darwin invented *natural selection*, allowing his adherents to salvage us from supposed religious superstition as we search the imperfect fossil records for evidence of an imperfect humanity that once rubbed elbows with our scaled and feathered friends. Darwin may well have been the father of using biology to supplant the humanities. Unlike Newton, who gave us the tools for knowing what idealized cannon balls in idealized vacuums will ideally do next, Darwin's theory is not physics; it does not predict what evolution will do next—ideally or otherwise. Instead, Darwin's genius was in borrowing science to persuade us on a view, not of science, but of history. Up until Darwin,[31] only historians and liars had the freedom to revise yesterday to fit the passions of today. Using science to revise history was one small step for history, one giant leap for the expansion of biomythological turf.

Charles Darwin—whatever the skeptic's doubts about his explanations of nature—was a rhetorical and networking genius. His *Origin of the Species* sold out on the day of publication. A century and a half later, *Origin* remains an astonishing literary success. A skeptic can't help but admire the achievement. Who could doubt the rhetorical devices that not only sold how nature selected the intricacies of the eye, but also eliminated any need for a window to the soul?

By the time Darwin wrote *Origin*, evolution was already old news. Old news is no news. No news is good news. Following the rules of logic, Darwin's skill at rhetoric turned old news into good news for those who would exclude purpose and value from the universe. Darwin is more than a superb author with a clever idea. He is a man whose idea has become a metaphor for biology replacing the humanities in capturing what it means to be human—a metaphor for science conquering transcendence and reducing hurricanes, sunsets, and loved ones' smiles to physics. Or so any good skeptic might suspect.

Just as Petrarch invented the Dark Ages to salvage the humanities from medieval scholars, and Darwin invented natural selection to salvage *us* from the humanities, I have reinvented the term *biomythology* to salvage skepticism from a growing number of scientific tales culling the stories of purpose, will, and transcendence from what it means to be human: a mythology rivaling, the skeptic may suppose, anything Greece, Rome, Petrarch, or Darwin could have imagined.

Biomythology

Biomythology may be characterized as the practice of borrowing science to change minds rather than matter. The biomythologist confuses the art of science with the fMRI's[32] ability to paint the human spirit by the numbers. Worse yet, to the mind of the skeptic, by inventing neuronal mythologies about *how* we think, biomythology tries to persuade us on *what* to think. Neuro-morality dictates values; neuro-theology reduces transcendence to the genuflecting of biochemicals.

The skeptic suspects that, to borrow Yogi Berra's calculus, biomythology is half reason, half experience, and 90 percent viewpoint. It's like a newly balding man's statistically significant, insightfully reasoned plea for a red sports car. Desire—whether to impress the girl or profit the pharmaceutical company—dictates both experimental design and interpretation.[33]

The skeptic also imagines that real science provides *mind-independent* truths about the physical world: the laws of aerodynamics lift your plane whether you are reading in the loo or sleeping in an overhead bin. Biomythology promises mind-independent truths about the *mind*—a task as daunting to the skeptic as finding a bachelor kissing his wife in a haystack. To the skeptic's doubtful way of thinking, not even an all-powerful God can create a mind-independent truth about the mind. Logic, language, Saint Thomas Aquinas—all forbid it. The skeptic may accept this summary: if real science is justified by its works, then biomythology is justified by our faith. Real science predicts; biomythology persuades. Real science astounds; biomythology enchants.

Scientism

If the skeptic were looking for the closest synonym to biomythology, it would be *scientism,* defined in *The Shorter Oxford English Dictionary* as "excessive belief in the power of scientific knowledge and techniques, or in the applicability of the methods of physical science to other fields, esp. human behaviour and the social sciences." Contemporary philosopher Thomas Nagel elaborates:

> Scientism ... puts one type of human understanding in charge of the universe and what can be said about it. At its most myopic it assumes that everything there is must be understandable by the employment of scientific

theories like those we have developed to date ... as if the present age were not just another in the series.[34]

The skeptic need not deny that psychology is an art. A gifted psychologist, whether inspired by the metaphors of science or the humanities, is a gifted artist who can profoundly change lives. Claiming that psychology deserves special attention compared to the humanities because it is a science, however, is scientism. The skeptic may find that term, while accurate enough, too cold, too rational; it merely illuminates. *Biomythology* inflames. Biomythology is scientism on a bad-hair day. The good skeptic may extend the term *biomythology* to all scientific explanations when used to engineer belief rather than matter—especially the beliefs of nonscientists in nonscientific settings.

Real science, like spirituality, is the search for truth; biomythology, like organized religion, provides the packaging. French poet Paul Valéry (1871-1945) captured for the skeptic the distinction between real science and biomythology when he wrote, "Science means simply the aggregate of all the recipes that are always successful. All the rest is literature."[35]

Often, bad literature.

The true difference between real science and biomythology is the difference between know-how and know-why. While many skeptics may doubt that know-how is proof of know-why—a mood elevated by amphetamines does not necessarily prove a "biochemical imbalance"—only biomythologists claim know-why in the complete absence of know-how. We know everything about nature's selections, except to predict what nature will select next. Know-why without know-how, the skeptic suspects, is like pretending a complete understanding of cancer with no clue for a cure, a complete understanding of passion with no lips for a kiss.

Warning

To conclude, from the skeptic's viewpoint biomythology advertised as real science is a bait and switch. To the skeptic, Darwin was no Newton. Advertising space travel and delivering natural selection is like advertising the Mona Lisa's smile and delivering a Rorschach test. The guillotine may have been the least pleasant legacy of the Enlightenment, but the skeptic guesses that selling ideology as science proves more persistent, pernicious, and perverse.[36] Consider a warning: unless your inner skeptic can instantly recognize when science is being used to engineer belief rather than the physical universe, a biomythologist's reason and evidence can be as convincing as a serial killer's smile on your first date.

Chapter 2

Exploring Biomythology

Having introduced the doubter to the terms *skepticism* and *biomythology*, we will continue to explore, expanding on our definitions.

Derivation

Biomythology derives from the Greek *bios*, meaning "life."

There is considerable debate about when life begins, but in science there has been even more debate about what life is. For the past 150 years, biologists, those whose science it is to study life, have been unable to agree upon a definition of what they are studying. Noam Lahav in *Biogenisis: Theories of Life's Origins* surveys the century and a half to collect fifty different proposed definitions of life.[37] In 1855, for instance, three years before Darwin came out of the evolutionary closet, German philosopher and physiologist Ludwig Buchner (1824-1899) was already avowing, "Spontaneous generation exists, and higher forms have gradually and slowly become developed from previously existing lower forms, always determined by the state of the earth, but without immediate influence of a higher power." In 1868, "Darwin's bulldog" Huxley, whom we have already met, concurred: "The vital forces are molecular forces." Harmonizing with this tradition of materialism, Friedrich Engels, who coauthored *The Communist Manifesto* with Karl Marx,[38] wrote in 1880, "No physiology is held to be scientific if it does not consider death an essential factor of life ... Life means dying." Stalin and Mao provided the research, confirming Engels' hypothesis with over seventy-five million case reports: life means dying.

Not all biologists, or skeptics, have agreed. Some have imagined that reducing life to materialism is like reducing the fire of sunset to a time of day, like trying to pour Lake Michigan into a shot glass, like trying to use a microscope to catch a rainbow. In 1855, Rudolf Virchow (1821-1908) wrote, "Life will always remain something apart, even if we should find out that it is mechanically aroused and propagated down to the minutest detail." In 1871, physician and microscopist Lionel Smith Beale concurred that "Life is a power, force, or property of a special and peculiar kind, temporarily influencing matter and its ordinary force, but entirely different from, and in no way correlated with any of these."

While life has been reduced to matter or a vital force exploiting matter, there is a third choice, a choice for skeptics and best put into words by twentieth century physicist Niels Bohr, who assured us that life was an elementary fact of nature. "Elementary fact" in scientific jargon is a synonym for "an abiding mystery":

> The existence of life must be considered as an elementary fact that cannot be explained, but must be taken as a starting point in biology, in a similar way as the existence of elementary particles forms the foundation of atomic physics.[39]

Why are there elementary particles? Why is there charge? Why is there space-time? Why is there life? How can three persons be one God? Ask the right questions, and you'll eventually collide with mystery, or as Huxley allowed, "fundamental incomprehensibilities." Matter, vital force, mystery—all have their defenders. Matter lovers declare any doctrine of vital force to be anathema. The skeptic may still have doubts. Has biochemistry's success in engineering pharmaceuticals beguiled us into rejecting vital force as a superstition of the past? Do antimicrobials create life or destroy it? Can biochemistry start life from scratch? Are we wise to abolish all vital forces from life until they are so not vital that we can start life without them? Even the cosmology of Richard Dawkins remains cautious: "The origin of life only had to happen once. We therefore can allow it to have been an extremely improbable event, many orders of magnitude more improbable than most people realize."[40]

Dawkins is suggesting that the original experiment may have been so difficult that it cannot be replicated, thus eliminating it from the realm of science. In other words, constructing life from matter may postdate the invention of the perpetual energy machine. Until all the data is in, the skeptic may insist that any insistence on matter possessing a monopoly on life is ideology, not science.

The New Vitalism

Is life more than a mystery? To help introduce his book *The Mysterious Flame: Conscious Minds in a Material World*, contemporary philosopher Collin McGinn shares the following riddle:

> This book is about the mystery of consciousness. My main theme is that consciousness is indeed a deep mystery, a phenomenon of nature on which we have virtually no theoretical grip. The reason for this mystery, I maintain, is that our intelligence is wrongly designed for understanding consciousness. Some aspects of nature are suited to our mode of intelligence, and science is the result; but others are not of the right form for our intelligence to get its teeth into, and then mystery is the result.[41]

McGinn's arguments suggest that we have about as much chance of solving the mystery of how consciousness springs from raw meat as a colorblind mystic has to perceive the hues of Joseph's Technicolor dream coat. Our senses and intellects are not "evolved" for the task. Thus, both mind and life escape our best efforts to solve their mysteries at this time, and as contemporary philosopher Alva Noë suggests, "it may be that the problem of mind and that of life are in an important sense one."[42] In other words, to the skeptic, trading consciousness for biology is trading one mystery for another.

Skeptics, of course, love mysteries in mysteries. We may know little with certainty, but, being unbound by convention, we are certainly free to imagine a new vitalism, a new metaphor, in which consciousness and life are inseparable. Defining life as "biochemistry

21

and consciousness," however, may be hard to sell until science develops an instrument for measuring consciousness. Knobs, gauges, and dial readings are a surefire way to promote new stories. If we could point a "consciousness meter" at an amoeba and find a dab of consciousness, the new vitalism would move from fairytale to science (and probably back to fairytale again upon losing a later vote of consensus). Research on the project has already begun: a neuroscientist has obtained an fMRI response on the still-living cells of a dead salmon.[43] The new vitalism cannot be far away.[44]

The idea of cells being conscious, of course, is an old one. Pierre Teilhard de Chardin (1881-1955), possibly one of the most influential paleontologists in the twentieth century,[45] went even further and suggested there could be "some sort of psyche in every corpuscle."[46] Although we are talking about minimal consciousness, not consciousness of world peace, Lahav did not quote Teilhard in our previously discussed definitions of life.

Even if skeptics embrace the biochemicals-and-consciousness story, they can safely doubt that the tale will go viral on bioscience listservs. The French Father of the Scientific Method, René Descartes (1596-1650), posited that "animals were non-sentient machines, devoid of a soul and indeed bereft of anything comparable to the rich inner mental lives of human beings. Animals, in the words of one of his sadistic disciples, 'eat without pleasure, cry without pain, grow without knowing it; they desire nothing, fear nothing, know nothing.'"[47] The good skeptic suspects that having enjoyed torturing animals for a number of centuries, biologists have about as much incentive to include consciousness in the definition of life as woodcutters have to include consciousness in the definition of *tree* or vegetarians have to include consciousness in the definition of *salad*.

NASA, Darwinism, and Life

Ignoring consciousness in the equation, biology's definitions of life continued to pour in until 1994, when NASA proposed one the skeptic may find worthy of the *bio* in biomythology: "Life is a self-sustained chemical system capable of undergoing Darwinian evolution." The definition is inspired. If Darwin's theory is incorrect, and the branches of the tree of life evolved in miraculous quantum leaps rather than gradual tiny steps, none of us will be here to despair or celebrate. Life, as NASA defines it, will simply cease to exist.

Does NASA's expertise in rockets and space qualify it as a spokesperson for life any more than Sophia Loren's acting qualified her as a spokesperson for spectacle frames? Is this misfiring of authority the reason why NASA's funding has not kept pace with its rockets? The skeptic wonders if, so long as life and consciousness remain but brute facts, brute facts will remain the facts of brutes.

Myth

Having left life to NASA, we turn the skeptic to the *myth* in biomythology. *Myth* is derived from the Greek *muthos* for "speech, myth, or fable." That speech is involved is hardly surprising to the skeptic. Think of some of the things that have come out of your own mouth. How many New Year's resolutions were, a week later, harder to find than a Greek god in a bingo parlor?

The Shorter Oxford English Dictionary defines myth as:

> A traditional story, either wholly or partially fictitious, providing an explanation for or embodying a popular idea concerning some natural or social phenomenon or some religious belief or ritual; spec. one involving supernatural persons, actions, or events; a similar newly created story.[48]

One of the skeptic's favorite myths is retold by Steven L. Goldman, the Andrew W. Mellon Distinguished Professor in the Humanities at Lehigh University, in his course *Science Wars: What Scientists Know and How They Know It*: "Space, according to Newton, is absolute, infinite, and uniform in all directions. Space is a featureless infinite container, existing totally independently of anything in space."[49] The story is "traditional"; Newton's absolute space reigned supreme for 200 years. The story is "partially fictitious"; the absolute part of absolute space had to be tossed. The story qualifies as a "social phenomenon"; for years, it was the talk of the town and the talk of science. However, is the story supernatural?

24

An absolute time and space certainly transcended any nature we know today, and Alexander Pope (1688-1744), one of the best-remembered English poets of his Enlightenment-torn age, wrote an epitaph intended to adorn Newton's tomb in Westminster Abbey:

Nature and Nature's laws lay hid in night:
God said, Let Newton be! and all was light.

In science and even popular culture, Newton's reputation approached the superhuman, and superhuman beings qualify—until history exposes their mortality—as supernatural. Newton's myth continues to provide the skeptic with a truth that works for the modest needs of most of the world, except for those who visit atoms, black holes, or the speed of light when entertaining the kids on the weekend. More importantly, Newton awakened a culture to the possibilities of science. The real truth of his myth was that it inspired hope that humans could become the controllers rather than the victims of the physical universe: a wonderful gift, an ecstasy in the making. The hope sustains us still. If the Martians do come, science will save us; if the AIDS virus mutates around the wrong corner, goes airborne, and gets religion, science will save us; if a hole in the ozone becomes a shortage of ozone in the hole, science will save us. Even the skeptic finds no comfort in doubting that we are protected by a benevolent science.

Science as Myth

Biomythologists are fond of reducing Genesis to a myth, but the skeptic is in good company when reducing the scientific pretensions to myth. Consider a quote attributed to University of Cambridge philosopher Ludwig Wittgenstein (1899–1951):

> In the day of the Enlightenment, science was rightly seen as being in the forefront of the struggle against religious mystification, superstition, and dogma. Today science has replaced religion as the source of authority of truth. Every source of truth must, in the nature of things, also be a source of falsehoods against which it must itself struggle. One great and barely recognized source of intellectual mythology in our age is science itself. The unmasking of scientific mythology (which is to be distinguished from scientific error) is one of the *tasks of philosophy*.[50]

Nor was Wittgenstein by himself in appreciating the mythical qualities of science. He was joined by several of the most famous philosophers of science of the twentieth century. Thomas Kuhn (1922–1996) in *The Structure of Scientific Revolutions* noted:

> Historians confront growing difficulties in distinguishing the "scientific" component of past observation and belief from what their predecessors had readily labeled "error" and "superstition." ... If these out-of-date beliefs are to be called myths, then myths can be produced by the same sort of methods and held for the same sorts of reasons that now lead to scientific knowledge.[51]

Another of the greatest twentieth-century philosophers of science, Sir Karl Popper (1902-1994), contended, "Science must begin with myths, and with the criticism of myths."[52] And, University of California at Berkeley philosopher Paul Feyerabend (1924-1994) went so far as to allow:

> Knowledge ... is not ... a gradual approach to the truth. It is rather an ever-increasing ocean of mutually incompatible alternatives, each single theory, each fairy-tale, each myth that is part of the collection forcing the other into greater articulation, and all of them contributing, via the process of competition, to the development of our consciousness. Nothing is ever settled.[53]

According to all four twentieth-century philosophers, science shares components with myth. According to Feyerabend, the disputes between Creationists and Darwinians are just part of the ongoing process, which develops our consciousness. Still, are all myths really equal? That, I suppose, depends on your perspective. The following is the skeptical perspective of this book.

Explanations

Science, like life, is composed of actions and stories. Biomythology is big on stories, short on action. The stories are also referred to as explanations. Explanations are tools, not truths. The tools are not necessarily interchangeable; they have their specific purposes. A hammer belongs in an explanation for driving a nail, not a china bowl. Gasoline belongs in an explanation about fueling

27

a car, not a heart. Borrowing real science to explain what makes us human is like applying the hammer to the bowl or soaking the heart in gasoline. The skeptic is free to doubt that the success with the nail or car predicts the success with the bowl or heart.

In the mind of the good skeptic, explanations simply satisfy our craving to believe. The skeptic hazards the guess that treating an explanation as truth is an act of faith that a better explanation will never be discovered in that deep time known as the future. Such treatment is also often a waste of time. Neither the hammer nor the bowl nor the gas nor the car benefits, even if the explanation is brilliant. An explanation's brilliance is a matter of aesthetics, not purpose—not the nail or gas. The brilliance of an explanation's logic is not truth, but merely a distraction from the search for truth. As the skeptic theorizes, Aristotle's explanations were brilliant; they distracted us from many truths for nearly two thousand years.

When it comes to explanations, the skeptic insists on those that predict the future, not the past. Let me explain. Suppose a husband gets home late, and his wife asks, "What took you so long, and why are your hands so dirty?" The husband provides an explanation:

> I finally got around to delivering your apple pie recipe to your sister. I was on Fourth Street zooming toward Pine Avenue when the light changed. I didn't want to get another ticket, so I slammed on the brakes, and the car stalled. When I turned the key to restart the engine, there was no sound—no clicking, nothing. You know how little I know about cars. I kept thinking

that someone would stop and help me. As always, that section of Fourth Street was deserted. I waited for fifteen minutes. Finally, this guy with Michigan plates drove slowly up and stopped.

At first, I wasn't sure I wanted his help. He was an odd-looking fellow with a wart almost the size of a grape on this chin. One of his earlobes was missing, as if someone had bitten it off in a fight. Looks aren't everything, though. He was really nice. He told me how to pop the hood, and we took a look. He pointed to the battery and then rolled his eyes as if I were a moron. Boy, did I feel stupid. The cable was not attached to the battery. I guess it must have been really loose, so it popped off when I slammed on the brakes. The guy told me to hang on for a moment. He opened his own trunk and rummaged around for about five minutes before he found what he was looking for: a pair of pliers. He didn't want to get his hands dirty, so he handed me the pliers and instructed me exactly how to put the cable back on the battery. It turned out the guy, Dan was his name, was lost. I couldn't remember all the street names needed to get him back to the freeway, so I drove ahead of him, taking him clear to the Cherry Street north onramp. I finally got to your sister's, grabbed the recipe to give to her, and smeared it with my dirty hands. We wasted another five minutes trying to read it, but the words right under 'six apples' were illegible. She asked if you could give her a call when you get a chance."

How brilliant is this explanation? It is chock full of facts; the facts logically fit together. The story accomplishes what a good explanation is designed to do: it pretty much ends the wife's curiosity and the need for future discussion. The sister and the fingerprints will be able to confirm the facts about the recipe, the smudge, and the identity of the smudger, should the need ever arise. Brilliant or not, the words tell us more about the cleverness of the author than about the truth of the explanation. This explanation, the skeptic would notice, has gaps precluding its full verification. The deserted street offered an excuse for a paucity of witnesses. We don't know the Samaritan's full name. He is from out of state, so his bitten ear and behemoth wart are unlikely to appear in the local fossil record. The Samaritan conveniently left no fingerprints of his own, no driver's license, and no forwarding address. The skeptic notes that the tale predicts no future, just excuses the past. It rambles, but does it capture in aesthetics what it lacks in parsimony?

The explanation may leave us skeptics skeptical, but like all explanations, it will be believed by all who want to believe it and rejected by all who don't. The wife, for instance, may prefer the explanation to discovering the truth: her husband's missing time in fact was passionately consumed in the arms of her sister; the dirty hands and ruined recipe were but a clever alibi concocted by the adeptness of the seasoned adulterers. The explanation's only prediction is that the husband is apt to rely on his dearth of mechanical ability to support future lies.

If the skeptic knows anything—which is doubtful—it is that Mother Nature, like the husband, is a philanderer and a liar, the multiplicity and duplicity of her trysts ever ready to seduce another scientist. The skeptic is not necessarily a doubter of past actions in history, but to the skeptic, the stories about what went down in the past are historical, not scientific. From the viewpoint of the skeptic, science plus assumption equals assumption. When it comes to explanations and assumptions about the past, the skeptic is careful to beware: the more credible the facts, the easier it is for the lies to close the case. For the skeptic, no case is ever closed—not by reason, not by evidence, not by facts, and not by wishful thinking about the power of reason, evidence, and facts. Mother Nature lies. To the skeptic's way of thinking, adding *scientific* to history leaves us with only history, if not with myth.

Myths, Magic, and Miracles

The skeptic is aware that real science provides myths, magic, and miracles; the myths inspire the magic to create the miracles. The skeptic also intuits that biomythology contents itself with myths, inspiring neither magic nor miracles. Real science has given us the biochemistry used to plagiarize the magical cocaine of the coco leaf, opium of the poppy flower, and natural mood elevator ephedrine—the model for amphetamine—from the Ma-Huang plant of Chinese medicine. Biomythology has given us the *Diagnostic and Statistical Manual of Mental Disorders* (DSM).

The skeptic notices that the drugs of real science oddly show little respect for the divisions of nature proposed by biomythology. Amphetamines affect ADHD and oppositional defiant disorder alike. Cocaine is likely, with a few exceptions, to raise moods across the full spectrum of the DSM. Thus, the "miracles" of the biochemistry occur with or without the mythological boundaries defined by biomythology. The words often aren't backed by "real" concepts in nature. In the words of philosopher David Lewis (1941-2001), "Properties carve reality at the joints—and everywhere else as well."[54] Until the skeptic knows exactly what causes cause, until the skeptic knows exactly what energizes energy, any scientific explanation invoking cause or energy as concepts of reality rather than as conveniences for conversation may still qualify as nonsense. To the good skeptic, the frailty of language is no reason to abandon the causes and energies learned by our common senses in our common environments. For the skeptic, however, such frailty may be a reason to suspect scientific explanations overflowing their boundaries to question common sense and common practices in common worlds.

Mythmakers

In real science, the equations matter more than the myths. Newton knew that the math of how gravity varies according to mass and distance was more powerful than a hypothesis of how he squeezed the effect at a distance from an apple. In biomythology, the skeptic discovers that the explanations are more important than the equations. An fMRI color reconstruction squeezed from statistics is worth a thousand words; an equation predicting how brains squeeze the response of sensation from the stimulus of light is nearly worthless; any hint of free will or inattention is enough to erase the equal sign between stimulus and response in the equations of perception.

That said, the skeptic should consider that some real scientists stress myth; some do not. Einstein fascinated us with his tales of footraces against light. The French mathematical physicist Joseph Fourier, in his 1822 *Analytical Theory of Heat,* provided equations that accurately described the behavior of heat—its flow and its conductivity. In Fourier's day, one popular myth described heat as a weightless fluid called "caloric"; another described heat as the increasingly rapid movement of particles comprising matter. Fourier was content with the magic of his equations. He needed no myth.[55]

Danish physicist Niels Bohr, like Newton and Fourier, offered math, not myth. Fellow physicist Ernest Schrodinger, in addition to adding his famous equation to Bohr's math, told a fabulously strange myth about a cat, neither alive nor dead, that is in a superimposition state, a wave of probability combining the best of both worlds, life and death—at least until touched by the magic wand of consciousness. As Bohr well knew, Schrodinger's cat was the myth and Schrodinger's equation the magic that would lead to the miracles. The skeptic would agree with Bohr.

Summary of Biomythology

The term *biomythology*, as used in this book to prepare skeptics for the world of science, is not the study of plants and animals on Mount Olympus. Rather, the term serves as a straw man the skeptic can use for doubting, misconceiving, and discrediting any argument within rhetorical range. Depending on what we are serving up as fodder for skepticism, here is a list of convenient definitions for biomythology:

- the myth that biology can replace the humanities in capturing what it means to be human
- the faith that the scientific method can do for minds what it did for matter
- scientism on a bad-hair day
- the use of science to change minds rather than matter
- the science of persuasion
- the stories of science told to inspire belief rather than technology
- the rhetoric of objectivity
- science altering worldviews rather than the world we view
- painting the human spirit by the numbers

Skeptic, take your pick and allow your rhetoric to burn the straw man of your choice.

Skeptical Conclusions

From the perspective of the skeptic, biomythology borrows the words of science for the purpose of persuasion, of engineering culture rather than nature. Biomythology pretends that what works for a bank shot on a billiard table can be extrapolated to a hundred trillion nerve connections rebounding in a skull during a first kiss. While real science has produced the transistor to unite the world, biomythology has fabricated the bell curve to divide it. Why do missiles rise while the soul sinks? Quite simply, the skeptic may hypothesize that missiles behave according to the laws of physics; the human spirit does not.

Biological Brimstone

Richard Dawkins assures us, "Science and technology have changed our world more in the past century than it changed in the previous hundred centuries."[56] Brown University biology professor Kenneth Miller cautions us, "If evolution is indeed the cornerstone of modern biology, how can America consider itself a modern scientific nation when a majority of its citizens reject that cornerstone as unsound?"[57]

Many have fled the hellfire and brimstone resounding from an occasional pulpit. The skeptic wonders why we should trade the threats of one sermon for another. Despite the hellfire and brimstone preached by the biomythologists, for the skeptic to doubt Darwin is not to languish in a purgatory without the fruits of real science. It is not to lose real biology and biochemistry and abandon antibiotics, a cure for cancer, or an end to Alzheimer's disease. It is not to lose the salvation of our iPads and iPhones. But failure to be skeptical of those who cite the miracles of real physical science to sell the metaphors of biomythology may cause us to risk losing the transcendence of our iSelves.

Chapter 3

Schools, Courts, and Imbeciles

In our first two chapters, we alerted skeptics to the possibility that biomythology borrows real science for the purpose of persuasion, of engineering culture rather than nature. Let's consider one example: our schools.

When it comes to applying the scientific method to the mind, a maxim applies: what children learn first, they learn best. Contemporary philosopher and college professor Daniel Dennett, for instance, is concerned not just with higher education but with lower education as well. He recommends a class on comparative religion for "our children" and goes so far as to suggest that "the role of the Catholic Church in spreading AIDS in Africa through its opposition to condoms" should "be part of the mandated curriculum for both public schools and home-schooling."[58] Not having Dennett's training in logic, I cannot tell if his argument is sound or noise, but I remain skeptical about indoctrinating homeschoolers to believe that AIDS is primarily spread by priests wearing cassocks instead of condoms.[59] In reality, the argument is not really about AIDS or priests or condoms. The argument is about ideology.

Materialism Versus Idealism

If a tree falls in the forest without a soul to hear it, does it make a sound? If a biochemical falls in the brain without a soul to hear it, does it make a thought? To the first question, at least, philosopher George Berkeley (1685-1753) might have answered, "God would have heard it." For Berkeley, "to exist is to be perceived."[60] He was a philosophical extremist arguing that the sound of the tree was an idea in the mind. Put another way, the shrill of the dog-whistle is in the dog, not the whistle. That's why the whistle perks the dog's ears and not our own. Berkeley was an idealist—idealism being, according to *The Oxford Dictionary of Philosophy*, "Any doctrine holding that reality is fundamentally mental in nature." We will return to the ideal in later essays.

Another example of philosophical extremism, to be found at the other end of the spectrum from idealism, is materialism—not the quest for desirable material objects such as larger houses or smaller waists, but what *The Shorter Oxford English Dictionary* defines as "the doctrine that nothing exists except matter and its movements and modifications. Also, the doctrine that consciousness and will are wholly due to the operation of material agencies."

Materialism has a long and illustrious history. Its earlier years are captured in *The Swerve: How the World Became Modern* by Harvard humanities professor Stephen Greenblatt. Greenblatt calls on his prodigious research, talent, and beautifully crafted prose to support his theory that the creation of modernity rested on the rediscovery of the poetry of Lucretius (99 BC–55 BC) and his *On the Nature of Things*. According to Greenblatt:

> Lucretius believed that the sun circled around the earth, and he argued that the sun's heat and size could hardly be much greater than are perceived by our senses. He thought that worms were spontaneously generated from the wet soil, explained lightning as seeds of fire expelled from heaven.[61]

Greenblatt's interpretation of Lucretius's views continues:

> The stuff of the universe ... is an infinite number of atoms, moving randomly through space, like dust motes in a sunbeam, colliding, hooking together, forming complex structures, breaking apart again, in a ceaseless process of creation and destruction.[62]

Lucretius's views were inspired by earlier philosophers such as Democritus (c. 460–370 BC) and Epicurus (341–270 BC). Greenblatt's thesis is that it was not mere philosophy, but Lucretius's poetry, that was responsible for perpetuating the materialist message that defines us as "modern": "Only the atoms are immortal."[63] Whether or not the skeptic believes that Lucretius was his age's poster child for poetic license, *The Swerve* mirrored the intellectual spirit of our times to capture both the National Book Award and the Pulitzer Prize.

Materialism Versus Naturalism

Natural philosophy, or science, as it is now called, is about nature, about naturalism, not about materialism. The terms *naturalism* and *materialism* are not to be confused. The two philosophies make radically different statements about reality. Naturalism is defined by *The Shorter Oxford English Dictionary* as "the belief that only natural (as opp. to supernatural or spiritual) laws and forces operate in the world." Compare this definition with Oxford's definition of materialism: "The doctrine that nothing exists except matter." There is a big difference between what *exists* and what *operates*. When I'm with my wife, for instance, my opinions exist far more often than they operate.

Naturalism makes no statements about transcendence, but rather centers itself on the control of nature by natural means. Put another way, naturalism limits what goes on *in* nature; materialism limits what goes on *outside* of nature. While skeptics insist we are blind to the full possibilities of mystery, materialists insist mystery is blind to us.

Possibly to dispel superstitions about what does and does not fall outside nature, one of the most influential philosophers of his day and ours, Ludwig Wittgenstein (1889-1951), erected a metaphysical firewall to protect transcendence from viruses and philosophers:

> Logic pervades the world: the limits of the world are also its limits. So we cannot say in logic, 'The world has this in it and this, but not that.' For that would appear to presuppose that we were excluding certain possibilities, and this cannot be the case, since it would require that logic should go beyond the limits of the world; for only in that way could it view those

limits from the other side as well. We cannot think what we cannot think; so what we cannot think we cannot say either.[64]

Does nothing fall outside the material world? Wittgenstein just placed transcendence beyond the limits of logic. Thus, materialism is, in a word, illogical—or so Wittgenstein, as I misinterpret him, might lead us to believe.

Nature and God

Naturalism is another matter. Ever since Bacon implored us to "seek everything from things themselves,"[65] naturalists have used explanations of nature and nature alone to control nature. Materialism is an attempt to control not only nature, but our stories about nature. Materialists want a monopoly on intellectual property—literature, science, philosophy, and even skepticism, claiming that skepticism about the supernatural is more holy than skepticism about the natural: the inspiration for this book! Skeptics remain skeptical about dividing mystery into categories. Mystery is mystery. The rating of mystery is like a beauty contest in which the judges are locked outside the auditorium.

Once Queen Elizabeth I dictated the religious rituals of her subjects, but not their religious beliefs. Her statement on the subject was, "I have no desire to make windows into men's souls." Materialism has no such compunctions. To confine its own speculations is not enough; it insists on confining the speculations of others, a practice apparently as much a part of the human condition as bedwetting, but unfortunately harder to outgrow. If biomythology, as I defined it, is the use of science to control our minds, then confiscating science to sell materialism fits the definition nicely.

How Does God Do It?

Whether or not God is welcome in science varies with the fashions, much like the width of lapels. In 1660, England's Academy of Science, The Royal Society of London for Improving Our Natural Knowledge was formed. Whereas before, philosophy concerned itself with all possible existence and knowledge, now there was a *natural* philosophy devoting itself to nature. Today some, enamored of science's credibility and prestige, imagine God needs to ride on the coattails of science, but once it was the other way around. Natural philosophy—before it had achieved any practical use—was sold to the public as a way of studying God by studying His Creation.

Sir Francis Bacon (1561-1626), whose contributions had largely inspired the Royal Society, wrote that the "Divine Mind" was revealed by the "true prints and signature made on the creation."[66] Robert Boyle (1627-1691), knowing full well where he got his gift for science, wrote in 1663, "yet perswaded I am, that the favor of God does (much more than most Men are aware of) vouchsafe to promote some Mens Proficiency in the study of Nature."[67]

Johannes Kepler, whose elliptical orbits of the planets saved us from Aristotle's fairytale circular orbits still clung to by Copernicus and Galileo, saw nature as "the clearer voice of God."[68] He saw those who studied nature as "priests of the most high God with respect to the book of nature"[69] and "God as geometry itself."[70] Sir Isaac Newton (1642-1727) similarly wrote with "an Eye upon such Principles as might work with considering Men, for the Belief of a Deity."[71] He believed that "the Motions which the Planets now have could not spring from any natural Cause alone, but were impressed by an intelligent Agent."[72] Despite the source of his inspiration, Newton became President of the Royal Society in 1703.

Today, Francis Collins, head of the Human Genome Project, continues this scientific tradition. In *The Language of God: A Scientist Presents Evidence for Belief*, Collins writes of the Human Genome as "God's instruction book." Like Bacon, Boyle, Kepler, and Newton, Collins is inspired by God. However, a caveat applies when combining God and science.

Rutherford divided science into physics and stamp collecting. I divide science into real science and biomythology. There is another way to divide science: discovery and verification. When it comes to discovery, anything goes. For the scientists mentioned in the above paragraphs, God was part of the equation, but God was not in the equations. Newton, as tradition has it, got hit in the head with an apple. Fleming discovered penicillin by accident, noting how bacteria had been killed in a contaminated petri dish.

August Kukule, credited with the discovery of the benzene ring, purportedly was inspired by his dream of a snake swallowing its tail. Einstein imagined footraces against beams of light. God, blunt trauma, accidents, dreams, imagination—all are allowed when it comes to scientific discovery. In science, the creative process runs unbridled, but when it comes to verifying the discoveries, the rules change. Then reason, repeatable observations, and the vote of scientific consensus stumble in. God and imagination may be as much as part of the scientific method as reason and observation. Science could be defined as "how God did it." "God did it," however, defines only faith.

Thomas Nagel—Naturalism without Materialism

Contemporary philosopher Thomas Nagel is a naturalist, but neither a theist nor a materialist. He is a neutral monist, to be exact. He argues there is only one substance composing all of nature. That substance includes both mind and matter. His philosophy eliminates the problem of life and consciousness arising from matter. Nagel's nature also includes purpose and value. His concern with materialism's attempt to take over naturalism and science is suggested by the title of his most recent book, *Mind and Cosmos: Why the Materialist Neo-Darwinian Conception of Nature is Almost Certainly Wrong.* He finds this conception "a heroic triumph of ideological theory over common sense."[73]

Materialism, not naturalism, claims nature is without purpose and value. For Aristotle, things had more than their matter, movers, and forms: they had their purposes, or—as the scholastics in the Middle Ages dubbed them—their "final causes." Teleology is the idea that the cosmos, with or without a God, may have a built-in goal or purpose or final cause. If our universe were expanding and contracting like an accordion, then teleology could be defined as the muscle memory of the universe. Today teleology is anathema among materialists and, as Nagel suggests, this ideology has captured the minds of contemporary naturalists. According to Nagel, naturalism does not exclude teleology, nor does teleology demand a belief in supernaturalism:

> The teleological hypothesis is that these things [forms of life, consciousness, genetic material] may be determined not merely by value-free chemistry and physics but also by something else, namely a cosmic predisposition to the formation of life, consciousness, and the value that is inseparable from them.[74]

The question of where the one substance came from persists, but there is no reason why that substance—known as nature—cannot be teleological. This barring of teleology from nature we owe in part to the disciples of Darwin. As philosopher and historian of science, Thomas Kuhn writes, "For many men the abolition of that teleological kind of evolution was the most significant and least palatable of Darwin's suggestions. The *Origin of Species* recognized no goal set by God or nature."[75] Thus, Kuhn spells it out. Teleology could have been dictated by either God or nature. Darwin, not naturalism, excluded both possibilities with his belief that natural selection resulted from mere competition between organisms for food and sex.

Despite the growing confusion, naturalism is not materialism. While few Americans believe in the metaphysics of materialism, most apparently respect and support naturalism. In 2009, for instance, the federal budget allotted $54,800 million for scientific research and $155 million to the National Endowment for the Arts. That we as taxpayers willingly spend over 300 times more on science than the arts hardly suggests that we are adverse to naturalism.

Perverting the Goal of Science

For the skeptic, naturalism remains a rule, not a goal of real science. Eager to "arrive at a knowledge highly useful in life,"[76] the French father of philosophy and science, René Descartes, reasoned:

> Knowing the force and action of fire, water, air, the stars, the heavens, and all the other bodies that surround us, as distinctly as we know the various crafts of our artisans, we might also apply them ... and thus render ourselves the lords and possessors of nature.

Across the channel, Bacon felt the goal of science was for man "to recover the right over nature." He wrote:

> In religion we are taught that faith is shown by works; and the same principle is well applied to a philosophy, that it be judged by its fruits and, if

> sterile, held useless; the more so if instead of fruits
> of the vine and the olive, it produces the thistles and
> thorns of disputes and controversies.[77]

What is more "sterile" than the predictions based on what nature may select next? What is more disputed and controversial? Bacon further insisted, "Nature, to be commanded, must be obeyed."[78]

While real scientists find ecstasy in rising above nature to control it, there are those who warp the history of real science to promote ideology. They promote Copernicus as "the father of the scientific revolution," not because of his inaccurate math, but because his theory contradicted the teachings of the Church. Copernicus's cosmos, like Galileo's, was an atavism of ancient Greek philosophy and Aristotle's circular orbits. It took Kepler and Newton to reconcile the math with the music of the spheres. In the same way, the materialists are obsessive about granting Darwin more space in high school science texts than is reserved for quantum mechanics. Quantum mechanics outshines Darwin's story in workable technology by a ratio of at least a thousand to one, allowing a third of our economy.[79] The skeptic suspects that Darwin's main economic contribution is in book rights. The teaching salaries and research grants of his disciples are paid, often in state-owned universities, by our taxes.

Some, nevertheless, continue to love and laud Darwin, not because his technology promises to fill their bellies or heat their homes, but because Darwin's rhetoric promises to fuel ideology. Take Edward O. Wilson's accolades:

> Charles Darwin's *On the Origin of Species* can fairly be ranked as the most important book ever written. Not the most widely read, to be sure. Copies of the *Origin* have not been placed in hotel rooms across America; its verities are not preached in pulpits each Sunday ... It is the masterpiece that first addressed the living world and ... humanity's place in it, without reference to any religion or ideology ... Its arguments have grown continuously in esteem as the best foundation for human self-understanding and the philosophical guide to human action.

Pay no attention to the eugenics and racism hiding behind the curtain.

Why place Darwin's book at the top of the scientific and philosophical heap? To hear Wilson tell the story, one might imagine that Darwin's principal contribution was to "guide human action" by replacing the preacher in the pulpit with the sermon of the sociobiologist. What happened to Bacon's demands for "works," his fervent wish for us to "recover our right over nature"? What happened to Descartes's plan to render us the "lords and possessors" of nature? If Darwin's disciples are correct, their prophet got it backwards. He recovered nature's right over us.

This urge to replace naturalism as a tool of science with materialism as the goal of science is nothing new. In an 1860 debate on Darwin's theory of evolution, T. H. Huxley (1825-1895) is reported to have responded to Bishop Samuel Wilberforce:

> I asserted—and I repeat—that a man has no reason to be ashamed of having an ape for his grandfather. If there were an ancestor whom I should feel shame in recalling it would rather be a man ... who ... plunges into scientific questions with which he has not real acquaintance, only to obscure them by an aimless rhetoric, and distract the attention of his hearers from *the real point at issue* by eloquent digressions and skilled appeals to religious prejudice [italics added].[80]

Were "the real point at issue" the advancement of scientific control of the physical world, then Huxley may have been correct. But, when "the real point at issue" is the prostitution of scientific explanation to establish an ideology in the larger culture rather than an advance in the control of nature, then science becomes biomythology and "aimless rhetoric" the order of the day. With aimless rhetoric in mind, we return to education.

School Days

While Dennett feels it is the duty of early education to teach "our children" about frocking, frolicking, and AIDs, our founding fathers were guilty of their own contributions to superstition. Consider these words by Thomas Jefferson:

> We hold these truths to be self-evident,
> that all men are created equal,
> that they are endowed by their Creator
> with certain inalienable rights.

Creator with a capital C—what was Jefferson thinking? Luckily, he neither impersonated a scientist nor appended his words to a public high school biology text. Had he, would the fervent protectors of the Constitution have burned the book, made lye soap from the ashes, and washed out the Founding Father's mouth? Bless my anachronisms, of course not! In 1776, sixty was not yet the new forty, nor "Creator" the new profanity. There was no constitution until 1789, no United States public high school until 1821, no word "scientist" until Cambridge master William Whewell coined the term to replace "natural philosopher" somewhere around 1840. It was a different world, a world in which political independence had yet to decay into political correctness.

Rhetoric, the use of language to persuade, continued to flourish, but in 1859, Charles Darwin turned persuasion into a science and wrote *On the Origin of Species by Means of Natural Selection, or the Preservation of Favoured Races in the Struggle for Life* (later abbreviated to *On the Origin of Species*).[81] Like Jefferson before him, Darwin wielded words to incite revolution—this time from Jefferson's Creator instead of from England.

The revolution continues. In today's schools, persuasion is hard at work. The good old Golden Rule has been replaced by the science of the good old metal detector. Is a change in indoctrination in order? Maybe we should train budding scientists, and those who will one day be forced to support them through tax dollars, with a good high school class on philosophy of science, team-taught by the ghosts of Thomas Kuhn, Paul Feyerabend, and Karl Popper. These

49

three famous twentieth-century philosophers, as we have seen, and will explore, refused to capitalize the T in science's Truth.

Miracles Versus Explanations

In science, formulas either work or they don't, at least for a given range of actions. Explanations cannot compete with equations. Words always remain suspect. For example, suppose in our superstition-free schools each time a male athlete set a record, the prom queen presented him with a rose and a warm embrace. Would the rise in new school records prove the power of roses? Similarly, did the miracles of Newton's math prove his myth of absolute space and time? Sure, the prom queen and Newton's equations were effective, but the explanations of roses and absolute space were hardly Truth. Why not teach students that the more absolute brand of science depends on equations, that culturally relative science depends on consensus, that aerodynamics remains the same over Massachusetts and North Carolina, and that postmodern evolution varies wherever better fossils are sold? Easier said than done.

Sticking It to High School Science Texts

In Cobb County, Georgia, where I live, the Board of Education affixed a sticker inside the cover of its high school science texts:

> This textbook contains material on evolution. Evolution is a theory, not a fact, regarding the origin of living things. This material should be approached with an open mind, studied carefully, and critically considered.[82]

The sticker is, of course, ridiculous. Evolution tells us absolutely nothing about the "origin of living things." In *Origin*, the best Darwin could offer was metaphor: life's "several powers, having been originally breathed into a few forms or one."[83] The theory doesn't suggest how to originate so much as a living cell from mere matter—we still haven't a clue. Still, what's the problem with the part about "an open mind"? Isn't this exactly what good science is about? Don't real scientists (the ones providing inspiration to engineer the physical world, not just the beliefs of the public) do exactly that?

Consider the recommendations of seventy-two Nobel Laureates, seventeen State Academies of Science, and seven other scientific organizations:

> Just as children should understand and appreciate the scientific theories that offer the most robust and reliable naturalistic explanation of the universe, children should also understand and appreciate the essentially tentative nature of science. In an ideal world, every science course would include repeated reminders that each theory presented to explain our observations of the universe carries this qualification: "as far as we know now, from examining the evidence available to us today."[84]

Why do these rules of scientific childrearing not apply to evolution? Why did the sticker's insistence on an open mind and critical consideration fall afoul of the constitution? Rather than singling out Darwin, should the sticker have said, "Take this entire science book with a grain of salt; the time bomb of discovery is always ticking, and when it explodes, it may blow away every *scientific* fact you just read, even if our *scientific* nuclear weapons don't"? Alternatively, is Darwinism not really science and therefore falls outside the protection of the Nobel Laureates' wisdom? My argument, of course, is fallacious. The sticker says absolutely nothing about Darwin, any more than it says anything about religion. I am creating a sticker that does not really exist and arguing against this imaginary sticker. In philosophy, this is known as the "straw man fallacy." In Cobb County, it's known as the *Georgia justice ad hominin fallacy*. In Georgia, we judge statements not on their contents but on the beliefs of their authors.

Reusable Conclusions Versus Facts

Had Nobel Laureate Richard Feynman written the sticker, it might have included, "And you have as much right as anyone else, upon hearing about the experiments (but we must listen to *all* the evidence), to judge whether a reusable conclusion has been arrived at."[85] Notice the use of "reusable conclusion," not *fact*. Sure, Darwinian apologists like Jerry Coyne maintain that "evolution is as solidly established as any scientific fact."[86] However, if science has taught us anything, then it is that facts change. Books have been written on that very subject. Consider these words from *The Half-Life of Facts: Why Everything We Know Has an Expiration Date*:

> Facts change all the time. Smoking has gone from doctor recommended to deadly. Meat used to be good for you, then bad to eat, then good again; now

52

it's a matter of opinion. The age at which women are told to get mammograms has increased. We used to think that the Earth was the center of the universe, and our planet has since been demoted.[87]

There are action facts and story facts. When I stub my toe, it hurts! Why I stubbed my toe or how it evolved are stories. To deify story facts is to demean the human potential for discovery or to tell a better story. Had Einstein, like a hundred years of scientists before him, accepted the fact that light was a wave and not inspired those who followed him, then quantum theory underlying the transistors and lasers of our digital age might have remained the stuff of science fiction.

The Central Organizing Principle of Biology

Coyne is not alone in his evolutionary enthusiasms. Michael Ruse and Joseph Travis, editors of *Evolution: The First Four Billion Years*, maintain, "The discovery of evolution is one of the greatest intellectual achievements of Western thought."[88] Evolutionist Theodosius Dobzhansky (1900-1975) believed, "Nothing in biology makes sense except in the light of evolution."[89] The father of sociobiology, Edward O. Wilson, opines, "So solidly have the fields of biology built upon the Darwinian conception of evolution that it makes sense today to recognize it as one of the ... universal principles ... that govern our understanding of life."[90] In *Only A Theory: Evolution and the Battle for America's Soul*, evolutionist

Kenneth Miller assures us that evolution is "the central organizing principle of the biological sciences."[91]

Where is skepticism when we need it? Once, life was the central organizing principle of biology; now, the primacy of life has been usurped by evolution. Apparently, no biologist can get along without it. Interestingly, in *The Double Helix*, the story of the most important biological discovery of the twentieth century, Nobel Laureate James Watson got along just fine without it; in the book's over 200 pages, I failed to find any mention of either Darwin or evolution. There are no illustrations of primal natural selection metaphysically chiseling the chemical bonds of the double helix.

Watson later made up for the omission. He left discovery behind to climb into the shopping cart of scientific consensus: "Evolution is the great unifying principle of all life, a law that underlies the history and the future of every species."[92] Watson had quite succumbed to Darwin's charms. In an October 14, 2007, *London Sunday Times* interview,[93] he confessed that he was "inherently gloomy about the prospect of Africa." Why? "All our social policies are based on the fact that their intelligence is the same as ours—whereas all the testing says not really." Watson received his Nobel Prize without mentioning Darwin. Ending Watson's credibility and career took the "central organizing principle" of biological ideology.

Instrumentalism Versus Realism

Whatever the truth of evolution and social Darwinism, you can always find philosophers of science who will agree that because scientific theories work they are facts, or true, or real. Such philosophers are called realists. Other philosophers and scientists, like Feynman above, claim that good theories are excellent instruments for controlling the world, but that the theories do not necessarily describe the *reality* of the world. These thinkers go by the name of instrumentalists. In our essay "The Road to Truth," we will cover realism and instrumentalism in more detail, but for now let's return to our high school science book sticker.

Scientific Truth is Decided by Scientists, Not by Judges

Should we be surprised that a judge in Georgia ruled that the philosophy of instrumentalism is unconstitutional? In Georgia, guns don't kill people; people kill people. In Georgia, ideas don't kill people; people kill people. But in Georgia we don't mess with a judicial system that has guns *and* ideas backing them. Nevertheless, as Coyne[94] assures us, "But scientific truth is decided by scientists, not by judges." The Oxford *Evolution: A Very Short Introduction* concurs, "The relentless application of the scientific method of inference from experiment and observation, without reference to religious or governmental authority, has completely transformed our view of our origins and relation to the universe."[95] We will say more on "inference from experiment and observation,"[96] but we agree; the scientific method is always sufficient to transform our views—except for when it isn't. Only then do we call in the judges. In the words of bad-boy philosopher of science Paul Feyerabend, (1924-1994), questioning the state interfering with the pursuit of knowledge:

> The courts of the Inquisition also examined and punished crimes concerning the production and the use of knowledge ... Today the much more modest wish of creationists to have their views taught in schools side by side with other and competing views runs into laws setting up a separation of church and state."[97]

Why do we have a separation between church and state? Was it to protect scientific consensus back when Pluto was still a planet? Probably not. The separation was designed so that Puritans can no longer hang Quakers over disputes in biblical interpretation[98]; so that a materialist, if elected President, could not tax religion out of existence; so that a New Age President could not enforce the New Age dogma banning all dogma. Separation of church and state is a blessing for us all. It allows our families openly—but not too openly—to practice their faiths no matter which religious or secular faction is in office. But, what does the separation of church and state have to do with instrumentalism versus realism? It seems that Georgia justice got it backwards. If anything, realism, with its claims of truth, is closer to religion than scientific skepticism. To tell children that Darwinism is *true* is far closer to faith than to tell them that Darwinism is an okay explanation—until human ingenuity triumphs and a new discovery or instrument allows us to come up with a prettier, more entertaining explanation.

Is my own book suitable for consumption in a public high school? I have been careful. I don't apologize for faith, except perhaps for faith in reason—an apology that remains necessary so long as cruelty is allowed to have its reasons. Since my writing is far from intelligently designed, it is my hope that it may find its way past Georgia justice. How then could the courts condemn me? Not even the scales of blind justice could be so blind. Could they?

How Many Imbeciles Does It Take to Change a Light Bulb?

In 1927, John H. Bell—a physician entrusted with the care of the feebleminded—wanted to sterilize eighteen-year-old Carrie Buck because she, her mother, and her baby daughter were all in danger of flunking their IQ tests and embarrassing the State of Virginia in the race to evolutionary utopia. The United States Supreme Court heard the case. Justice Oliver Wendell Holmes, Jr.—unwilling to allow the unintelligent to swim in America's gene pool—famously ruled, "Three generations of imbeciles are enough." Justice hardly gets any more just.

We may forgive Holmes. He was merely learning disabled—disabled from learning too much biomythology masquerading as real science. He badly needed this instruction manual—and so do the justices who, tempted by the beauty of those Technicolor statistics known as neuroimaging, even now continue to consider biomythology as evidence. For instance, when seventeen-year-old Christopher Simmons and an accomplice "wrapped a woman's head with duct tape, bound her limbs with electrical wire, and threw her off a railway bridge," he escaped the death penalty. As it turns out, Christopher did not murder the woman; his underdeveloped brain did. Quite possibly influenced by biomythological research on the "immaturity of adolescent brains," Justice Anthony Kennedy wrote the majority opinion, citing the "scientific and sociological studies" as confirming a "lack of maturity ... found in youth more often than adults."[99] A pretty safe opinion.

As we will see in a later essay, those who accept the survival of survivors can hardly reject the immaturity of the immature. Had Kennedy rejected the death penalty out of "respect for human life," rather than respect for brain science, he might have been impeached for using the wrong metaphor. Only time will tell if the maturing of the boy's brain will produce an exemplary citizen or a master duct-tape artist. Still, one wonders why the average eight-year-old, whose brain is obviously less developed, less mature than seventeen-year-old Christopher's, shuns duct-tape work, but who is to question biomythology?

Darwin's disciples, whether created or selected, have now been proselytizing for *six generations*. How time flies! Just imagine six generations of Darwinian disciples! This must be some kind of record. They claim we should let eugenic bygones be bygones; it can't happen again; we are too enlightened, having substituted genetic testing for intelligence tests to decide who should be admitted into our brave new world. We should forgive and forget past eugenic faux pas, even as we teach natural selection to prepare young minds to embrace the next experiment to enhance our racial purity and perfect humanity. Still, I could hardly blame school boards for hesitating to allow my safely impenetrable rhetoric into English classes. No skeptic would want to chance a court case in which a Georgia justice expanded the precedent of *Buck v. Bell* to read, "*Six* generations of imbeciles are *not* enough."[100]

Chapter 4

Intelligent Designs

How can we capture the heart of a culture? In the scientific literature, as in literature in general, the real answer is often, as every good skeptic should know, provided by reading between the lines, ferreting out the intelligent designs of the author from those of nature and history. Such literary criticism need not limit itself to exploring a particular section of the bookstore. Fiction, nonfiction—it makes no difference. We never know when science fiction will evolve into science, and science will evolve into science fiction. Not even natural selection can predict in which direction authority and consensus will change. In this essay, we will therefore read between the lines to explore whether natural selection tells us more about intelligent design or nature.

Pitching Paley

Do our current perceptions of nature reflect the designs of God or man? Imagine—as Christian apologist and moral philosopher William Paley (1745-1805) imagined in *Natural Theology, or Evidences of the Existence and Attributes of the Deity, Collected from the Appearances of Nature*—that you were strolling the heath and pitched your foot against a stone. Could you be faulted for assuming the stone's position was accidental, a chance product accurately representing the whims of nature? However, suppose that next to the stone you found a book, its pages swollen from humidity and gritted with dirt, its cover faded from age and sun, only a few words from its title page still visible: *"Natural Selection," "Preservation of Favored Races,"* "Charles Darwin."

Would you be equally justified to imagine that the words of the book reflected nature as solidly and accurately and enduringly as the stone? Judging from the evidence, would you attribute greater intelligence to the designer of the stone or the gritty pages?

Imagine that you opened the book to find Paley's personal God replaced with a personal Mother Nature, a natural selection that "is daily and hourly scrutinising [*sic*], throughout the world, every variation, even the slightest; rejecting that which is bad, preserving and adding up all that is good."[101] Still turning pages, you find, "When we see leaf-eating insects green, and bark-feeders mottled-grey ... we must believe that these tints are of service to these birds and insects in preserving them from danger."[102] Is this Mother Nature speaking or Charles Darwin?

Pitching Adam Smith

Imagine that it was your lucky day, and you found another dirty book on the heath—*Wealth of Nations* by the famous economist Adam Smith. Suppose the book's gritty premise was that individuals working for their own good produce the unexpected end of a greater good for the society as a whole: "He [the individual entrepreneur] intends only his own gain, and he is in this, as in many other cases, led by an invisible hand to promote an end which was no part of his intention." Now imagine that in comparing the two books, you noticed a similarity too great for chance between Darwin's natural selection and Smith's invisible hand. And you couldn't help but conclude that Darwin had merely turned Mother Nature into a *laissez-faire* capitalist by suggesting that in biology, like in economics, individual organisms competed for limited resources, the profits being paid in "reproductive success."[103]

Pitching Evolution

So ask yourself: was the picture of war, famine, and death, of organismic capitalism, creating the breadth and grandeur of life, intelligently designed by nature, or by Charles Darwin? For the skeptic, such questions never go away. Indeed, discussing evolution has always been precarious. In the middle of the 1700s, for instance, the Comte de Buffon, author of the monumental four-volume *Natural History*, hinted at evolution. He was soundly and enthusiastically ignored. In 1809, Jean Baptiste Lamarck's book *Zoological Philosophy* openly detailed evolution. Lamarck was ridiculed first from behind his back and then from above his pauper's grave.

In 1844, *Vestiges of Natural History of Creation* appeared in England. The book was a pioneering work in a genre that continues to this day to amuse succeeding generations with the latest versions of scientific truth tailored to fit our most popular pretensions, condescensions, and predilections—not to mention our love for a novel tale. *Vestiges* manufactured an evolution dictated not by God, but by the universe running on automatic pilot, in the deist fashion, according to the *natural laws*. The book was "roundly condemned on all sides" by everyone except those labeled as "freethinkers, atheists, infidels, and socialists"[104]: the skeptic suspects, pretty much the same crowd who today continue to keep the genre afloat. The book's author wisely remained anonymous.

So how did Charles Darwin evade the ostracism normally reserved for those promoting evolution? Having undergone the conversion process himself—albeit in the closet—he knew how to convert others. It all began years earlier when Darwin, demonstrating little interest or aptitude for the study of medicine at Edinburgh, transferred to Cambridge, where, as a sworn member of the Church of England, he completed the undergraduate degree that was a prerequisite for studying to become a member of the clergy. Darwin's Cambridge final exam, while hardly guaranteeing Darwin's competency as a naturalist, covered both William Paley's *Evidences of Christianity* and *Principles of Moral and Political Philosophy*. Darwin, however, preferred Paley's 1802 *Natural Theology*. Darwin later admitted, "I do not think I hardly ever admired a book more ... I could almost formerly have said it by heart."[105]

Pitching Malthus

Darwin admired other writers as well. In the eighth paragraph of *Origin*'s introduction, Darwin shares the source of his theory: "This is the doctrine of Malthus, applied to the whole animal and vegetable kingdoms." Thomas Malthus (1766–1834) was a cleric who wrote at the end of the eighteenth century, following in the footsteps of essayist Jonathan Swift (1667–1745), who in 1729 had authored "A Modest Proposal for Preventing the Children of Poor People in Ireland, from Being a Burthen on their Parents or the Country, and for Making Them Beneficial to the Public." Swift's "cheap, easy, and effectual" solution was for the Irish to eat their babies. He had learned from "a very knowing American ... that a young healthy child, well nursed, is at a year old a most delicious, nourishing, and wholesome food, whether stewed, roasted, baked, or boiled." The practice, Swift proposed, would clear "the streets, the roads and cabin-doors crowded with beggars"; provide "artificially dressed" skin of children to "make admirable gloves for ladies, and summer boots for fine gentlemen"; and make the children "sound useful members of the commonwealth."

Who could argue?

Malthus, coming along far too late to share Swift's priority, made a no less modest proposal of his own in 1798, *An Essay on the Principle of Population, as It Affects the Future Improvement of Society* ... [106] Malthus, "actuated solely by a love of truth," reasoned that, "Population when unchecked increases in a geometrical ratio.

Subsistence increases only in an arithmetical ratio." Which means that population will grow far faster than the food supply. Malthus wrote, "That population cannot increase without the means of subsistence is a proposition so evident that it needs no illustration."

So Malthus gave no statistics, nothing but his own self-satisfied if hardly self-evident opinion. He argued, "The poor laws of England tend to ... increase population without increasing the food for its support." Worse yet, the feeding of the poor was wasted "upon a part of the society that cannot in general be considered as the most valuable part." Reading well between the lines of scripture, the good cleric's implied solution was to starve the poor so that their deaths would maintain a proper equilibrium between population and subsistence and reduce overall suffering.

Swift aimed at irony as he satirized the callousness and arrogance of the English toward the Irish poor. Malthus wrote in all seriousness, providing the perfect excuse for not sharing his Christmas goose.

In 1838, Darwin read Malthus's idea, put two and two together and got three, continued to add up his thoughts with notes from his voyage on the *Beagle,* and finally came up with four: species had not been individually created; they had evolved through competition for survival. From this point, Darwin slowly began to waver from his orthodox faith in natural theology.

His apostasy was almost as gradual as he imagined evolution to be. Considering the social position of the Darwins and the reception that evolution generally enjoyed, Charles was in no hurry to join the casualty list. True, his grandfather had written on the subject, but Erasmus Darwin was a poet, not the respected author of books on exotic world voyages, coral reefs, and barnacles. Unshielded by poetic license, the younger Darwin's reputation was at stake. He kept his mouth shut ... and shut ... and shut.

Pitching Wallace

Then, twenty years later, in 1858, he received, in the mail, a curious manuscript: *On the Tendency of Varieties to Depart Indefinitely from the Original Type*. The scientific paper was about the "struggle for existence"; it dismissed Lamarck's idea that giraffes had developed long necks from stretching to eat leaves; it proposed a novel mechanism for evolution:

> Neither did the giraffe acquire its long neck by desiring to reach the foliage of the more lofty shrubs, and constantly stretching its neck for the purpose, but because any varieties which occurred among its antitypes with a longer neck than usual at once secured a fresh range of pasture over the same ground as their shorter-necked companions, and on the first scarcity of food were thereby enabled to outlive them. Even the peculiar colours of many animals, especially insects, so closely resembling the soil or the leaves or the trunks on which they habitually reside, are explained on the same principle; for though in the course of ages varieties of many tints may have occurred, yet those races having colours best adapted to concealment from their enemies would inevitably survive the longest.

Green leaf-eating insects and mottled-grey bark-feeders: Darwin had been scooped! The manuscript was essentially Darwin's theory of organismic capitalism authored by the self-taught naturalist, Alfred Russel Wallace, a fellow admirer of Malthus. While Darwin, having endured five years of seasickness, was more than glad to remain standing firmly on English soil, Wallace had been exploring the Malay Archipelago—those islands strung between Southeast Asia and Australia. Wallace had not been educated at Cambridge. Instead, he had attended Herford Grammar School. Unlike Darwin, who could afford to cultivate friends and supporters by donating to them his specimens, Wallace was obliged to sell his own collected specimens to make a living. Despite Wallace's acute interest in nature, he was regarded by better-educated scientists as a "mere collector."[107]

Wallace, as Darwin's friends warned, had been overtaking Darwin for several years. In 1855, the younger man's paper, *On the Law Which Has Regulated the Introduction of New Species*, was published to the acclaim of none. Samuel Stevens, the agent responsible for selling the specimens Wallace collected, "reported that the only reaction in the Entomological Society was grumblings that [Wallace] should stop theorizing and get on with his collecting." [108]

Wallace's earlier paper was also ahead of Darwin's published efforts: "Every species has come into existence coincident both in space and time with a pre-existing closely allied species." The "many-branched line" of the process is "as intricate as the twigs of a gnarled oak." Wallace proposed that the separate species on the Galapagos Islands were all "modified prototypes" of "the same species" that had earlier arrived on currents from South America. Thus, as early as 1855, Wallace was already in print hypothesizing such ideas as the common ancestry of current species, the tree of life, and the divergence of species from an earlier common species on the Galapagos Islands.

A perusal of the paper that Wallace sent to Darwin in 1858 shows that the two naturalists had their disagreements on the exact mechanism of natural selection. Wallace—a bachelor starving in the tropics—centered his theory on food. Darwin—well on his way to siring ten children—centered his theory around sex. Despite their quite distinct proclivities, Darwin was forced to act or lose to Wallace credit for the theory. After twenty years of sick worry, Darwin publically jettisoned Paley's Creator for the theory of organismic capitalism. Still smarting from the throes of his own conversion, Darwin took rhetorical pains to get his readers through the process in a less protracted and traumatic manner. From personal experience, he knew exactly what tiny leaplets of faith the meekest of the meek might require. He therefore intelligently designed his book to provide steps as tiny and gradual as a tiptoeing natural selection.

Pitching Darwin's Priority

Nevertheless, before he could sell his theory, his initial rhetorical challenge was to sell his priority—the fact that he, not Wallace, had discovered the theory. The *Origin* begins with the sentence, "When on board *H.M.S. Beagle,* as naturalist, I was much struck with certain facts in the distribution of South America, and in the geological relation of the present to the past inhabitants of that continent."

What was Darwin telling us? First, he was reminding us that he was the same renowned author who had written the renowned travel book about the *Beagle*'s voyage. Second, he was stressing that he was a professional naturalist even though his Cambridge final exams had centered on Homer, Virgil, Euclid, Paley, and Locke.[109] Third, even though Wallace's 1855 paper had already told the story of inhabitants from South America having been "modified" to form different species, Darwin assures us that he had already been there and done that, twenty years before; he had just neglected to mention it.

Darwin continues the first paragraph, telling the story of how the origin of the species had begun to occur to him in 1837 and from then on how he had begun "patiently accumulating and reflecting on all sorts of facts." He concludes by documenting his priority:

> After five years' work I allowed myself to speculate on the subject, and drew up some short notes: these

I enlarged in 1844 into a sketch of the conclusions, which then seemed to me probable; from that period to the present day I have steadily pursued that same object. I hope that I may be excused for entering on these personal details, as I give them to show that I have not been hasty in coming to a decision.

The paragraph is careful to establish that natural selection was Darwin's idea two decades before Wallace stumbled upon the same theory. Darwin downplays Wallace's courage, certainty, and perhaps foolishness, in pursuing publication. We read Darwin's explanation for not publishing sooner and come away, as intended, with respect for his caution and deliberation before revealing his theory. Darwin wasn't fearful of being ridiculed or ostracized; he had simply postponed sharing his revelations because he was a meticulous researcher, intent on getting the details exactly right. At least that is what he implied and history inferred. Moreover, Darwin impresses us with his modesty as he excuses himself "for entering on these personal details." Before further chronicling the priority, let's focus for a moment on Darwin's modesty.

Pitching Darwin's Modesty

"Friends, Romans, and countrymen." So Mark Antony saluted the mob before deconstructing Caesar's ambition and Brutus's honor. Antony would doubtlessly have had fun with Darwin's modesty as well, for as everyone knows, Darwin was a modest man. We read of his diffidence by all who have been sold by Darwin's rhetoric. David

Quammen, editor of the illustrated *On the Origin of the Species,* confirms that "Charles Darwin was himself a modest and affable man—shy in demeanor though confident of his ideas—who meant to persuade, not to declaim or intimidate."[110] Biographer Paul Johnson agrees:

> No scientific innovator has ever taken more trouble to smooth the way for lay readers without descending into vulgarity. What is almost miraculous about ... [*Origin*] is Darwin's generosity in sharing his thought processes, his lack of condescension. There is no talking down, but no hauteur, either. It is a gentlemanly book.[111]

Darwin was most certainly a modest man, although maybe not so modest as to be averse to fame. As Quammen allows, *"On the Origin of Species* wasn't written for experts. It was written for everybody who reads, thinks, and wonders."[112] Darwin's book was intelligently designed to advance both nature and culture—an achievement. Moreover, how modest were Darwin's ambitions for that nature and culture? Consider a confession found in the tenth paragraph of *Origin's* Introduction:

> Who can explain why one species ranges widely and is very numerous, and why another allied species has a narrow range and is rare? Yet these relations are of the highest importance, for they determine the present welfare, and, as I believe, the future success and modification of every inhabitant of this world.

Is explaining the "future success and modification of every inhabitant of this world" the modest goal of a modest man? Should we be skeptical, or am I taking this entirely out of context? Did Darwin not say what he said? Did Darwin suppose that science, once equipped with this epiphany, would continue to defer to nature on deciding the relative success of inhabitants? Darwin's modesty aside, we will return to how Darwin and his friends secured half of the priority for the theory that would soon be promulgated by eugenicists in their bid to determine the future success and modification of every inhabitant of this world.

Pitching the Priority

Wallace sending his memoir to Darwin was one of the greatest ironies of the history of science, but the incident nevertheless shows us of what intelligent designs the scientific method is capable. As a result of Wallace's blunder, Darwin came away with half of the priority, rather than no priority, for the theory of natural selection. His guilt is almost palpable as he offers the excuse "my health is far from strong." He finishes the narrative without assuming responsibility for what transpired:

> Mr. Wallace, who is now studying the natural history of the Malay archipelago, has arrived at almost exactly the same general conclusions that I have on the origin of species. Last year he sent to me a memoir on this subject, with a request that I would forward it to Sir Charles Lyell, who sent it to the Linnean Society, and it is published in the third volume of the Journal of the Society. Sir C. Lyell and Dr. Hooker, who both knew of my work—the latter having read my sketch of 1844—honoured me by thinking it advisable to

publish, with Mr. Wallace's excellent memoir, some brief extracts from my manuscripts.

Take note that Wallace had not submitted a scientific theory; he had sent a memoir of his "general conclusions." Darwin, perhaps not at his best after nearly losing credit for his life's work, continues to justify his actions by documenting in his scientific way that Hooker had read Darwin's 1844 manuscript. The paragraph also stresses that it was Hooker and Lyell, not Darwin suffering from weak health, who insisted Darwin share half of Wallace's priority.

In truth, had Wallace sent his manuscript elsewhere, he might have received full credit for his theory, but what good would it have done him? Without Darwin's name, status, colleagues, and skills at rhetoric backing him, Wallace would have remained an unknown specimen peddler.

Pitching Science

The whole affair calls to mind a Feyerabend discussion of the way science actually progresses:

> It has emerged that ... ignorance, pigheadedness, reliance on prejudice, lying, far from impeding the forward march of knowledge may actually aid it, and that the traditional virtues of precision, consistency, 'honesty,' respect for facts, maximum knowledge under given circumstances, if practiced with determination, may bring it to a standstill."[113]

To have given Wallace the priority he deserved might have done exactly that—brought the theory to a standstill. Even with Darwin's name attached, the Linnean Society reading created no stir. It was not the theory or the shared priority, but Darwin's networking and his rhetorical masterpiece that made natural history. The same author who had sold us on an intelligently designed "priority" sold us on an equally intelligently designed theory.

Pitching the Good News

The incident is interesting not so much for what it tells us about Darwin's ability to conceal himself behind his rhetoric, but for what it tells the skeptic about the scientific method and the necessity of reading between the lines of the *scientific* literature—something we can't accomplish by merely reading abstracts. Darwin presented objective fact after objective fact to assuage his guilt, justify his decisions, and establish his priority. Is it fair to suggest that the meticulous defense of Darwin's priority and theory reflected the intelligent designs of Darwin and his friends rather than natural history? Probably not, but Darwin's rhetoric most certainly opened the door for the modest ambitions of his theory to take hold.

Pitching Eugenics

Darwin's cousin Francis Galton (1822-1911) had intelligent designs of his own, but unlike Darwin, Galton displayed no modesty, false or otherwise. He didn't hide his ambitions for science to step in and, in the words of his cousin, control "the future success and modification of every inhabitant of this world." As historian Frederick Gregory recounts in his *Natural Science in Western History*:

> Dalton believed that it was the duty of democracy to defend itself against the introduction of 'degenerate stock' and to positively encourage marriages among those of good stock. He had attempted to identify, using statistical data, the qualities of the very best members of society in his book, *Hereditary Genius,* in 1869.[114]

Galton called his better-bell-curve-through-better-breeding philosophy "eugenics." This innovation, promising to control the relationships that would modify the future inhabitants of the world, captured the imagination of a public fascinated with all things progressive and *scientific,* or as G. K. Chesterton would later dub it in his 1922 book *Eugenics and Other Evils* (1922), "a modern craze for scientific officialism." Chesterton also wrote about the "pedantic quackeries" of his age: "It was a time when this theme was the topic of the hour; when eugenic babies ... sprawled all over the illustrated papers; when the evolutionary fancy of Nietzsche was the new cry among the intellectuals."[115]

Chesterton sounded the alarm and was soundly ignored. How could he criticize progress? How could he miss the excitement of the new science? It was no different than if today someone were to caution against the craze of dressing up words with "neuro"— "neuroscience" and "neuropsychology" and "neurolinguistics" and "neuro-education" and "neuro-theology" and "neuro-morality"—to sell them as science. Eugenics was no less of a craze in the early twentieth century.

By the 1920s, fifteen states in America had passed sterilization laws to naturally select against pollution in the gene pool. Germany appropriated the good old American knowhow of sterilization and supplemented it with the bad old Nazi knowhow of euthanasia—and eventually genocide—to work toward the "modification of every inhabitant of this world" inspired by Darwin's modest proposal. Today's Darwinians disavow eugenics, claiming that it was no more real science than burning witches was real Christianity. The only difference is that while witch burnings have abated, eugenics is flourishing.

Pitching Prenatal Eugenics

Today the intelligent designs of Darwin's disciples have been revived under the label of "prenatal genetic testing," a euphemism for prenatal eugenics. Who should be allowed to enter our brave new world? One day soon, we will be able to say goodbye to Down's syndrome, dyslexia, and blond hair. Thanks to science, we will no longer have to suffer the laughter of children born on the wrong side of the bell curve.

Pitching Biological Original Sin

To close our exploration of Darwin's intelligent designs, we will consider a quote selected by the editor, James Watson of double-helix fame, in his *Darwin: The Indelible Stamp, The Evolution of an Idea.*

Epigraph[116]

We must acknowledge, as it seems to me, that man with all his noble qualities, with sympathy which feels for the most debased, with benevolence which extends not only to other men but to the humblest living creature, with his god-like intellect which had penetrated into the movements and constitution of the solar system—with all these exalted powers. Man still bears in his bodily frame the indelible stamp of his lowly origin.

> —Charles Darwin
> *The Descent of Man*

Darwin indelibly stamped a new story into the minds of a culture. Our original sin hails, it seems, from the garden of biology, not Eden. It is only natural if we occasionally descend beneath our noble qualities—our sympathy, our benevolence, our godlike intellect—long enough to provide "for the future success and modification of every inhabitant of this world." Even as we descend, however, the skeptic must continue to ponder if we are following nature's, or Darwin's, intelligent designs.

Chapter 5

Evolutionary Equivocation

Having practiced our skepticism on reading between the lines, we turn to another of the devices of natural rhetoric in the art of cultural engineering: equivocation.

At Oxford, equivocation is the "use of ambiguous words or expressions in order to mislead." In science in general, preserving a theory is as simple as redefining all its terms to fit this year's discoveries. Thus the words of the theory stay the same, and the meaning evolves to conform to the passions of the times. For instance, as we will examine in a later chapter, *uncuttable* critters can be, without a name change, diced until the body parts fill an entire zoo. In this chapter, we will alert the skeptic to recognize and explore not just any equivocation but the rhetorical device of *evolutionary equivocation*.

When we claim that evolution is true, what exactly are we claiming? Is evolution change over time? Is evolution survival of the fittest? Is evolution Darwin's arrival of the fittest, in which tiny biomechanical advantageous variations allow organisms to compete for food and sex, and these advantages adapt different species to different environments and very gradually add up to not only new species, but entirely new branches of the tree of life? Or is the arrival of the fittest, the new limbs on the tree, formed by an evolution in which tiny internally dictated changes create profound leaps in development, defying Darwin's theory? Moreover, when we are arguing, can we vary the definitions as necessary to win the debate? The term *evolution* is a rhetorician's dream, providing an opportunity for an equivocation inside an equivocation inside an equivocation. The skeptic may delight in how well evolution conforms to Malthus's thinking: the theories grow arithmetically while the equivocations grow geometrically.

Political Definition

What is evolution? Definitions vary. To avoid controversy, the politicians of evolutionary theory equivocate, "Evolution is the theory that life on earth has gradually changed over time." How can we argue? Every time a species goes extinct, life on our planet changes—irrevocably and forever, or at least until it's time for the next *Jurassic Park* sequel. Bengal tigers and American eagles balance on the cusp of evolution as nature selects humanity over majesty and grace.

As any skeptic should understand, when the politic definition is invoked, it's not religious conservatives who oppose evolution; it's those bleeding-heart conservationists, who would delight in stopping evolution dead in its tracks. Consider a bumper sticker, "Save the Whales." The sticker might as well say, "Ban Evolution." Consider a more hopeful T-shirt, "Nuke the Whales." Now, that slogan conveys evolutionary enthusiasm at its best. Nor am I alone in my assessment. Contemporary author Stephen Cave, reflecting on the gratitude we owe to evolution, assures us, "A long, long history of death has made possible the incredible fact that you are alive now."[117]

Incredible.

Financing Evolution: Donations or Taxes?

May the skeptic wonder if the "evolution is the theory that life on earth has gradually changed over time" definition is merely a political contrivance? Can those who assert that "evolution is true" or "evolution is a fact" equivocate with impunity? If called on their assertions, the asserters need merely cite the politic definition.

But why the politics?

As the skeptic may suspect, those packaging evolution for public consumption go to great lengths to minimize controversy. And well they should. Richard Dawkins berates "the sums of tax-free money sucked in by churches,"[118] but in America, religion is financed by donation, not enforced taxation—the primary funding for the research and promulgation of evolutionary theory. Donations are voluntary; taxes are not. As a recent *Nature* editorial, "Science in the Schools," puts it, promulgating evolution is necessary "to ensure that the supply of competent young researchers and policymakers does not fail."[119] In other words, we don't want talented youth defecting to the Intelligent Design camp, and since policymakers divvy up the taxes and decide on funding, we certainly want them on the side of Darwin.

It may be skeptical to say it, but continuing the flow of tax dollars to Darwin is a real concern. According to a 2008 Gallup poll, also quoted by Richard Dawkins,[120] an evolution in which God plays no part fails to represent the opinion of at least eighty percent of Americans. Lest taxation without representation provoke another revolution, politically inspired equivocation for the good of the larger public becomes a necessity.

Equivocation Evolves

Despite the peril that expanded explanations pose to funding, definitions of evolution tend to be more involved than the politically correct version. Consider the actions of evolution offered by the 2007 edition of *The Shorter Oxford English Dictionary*:

> Any process of gradual change occurring in something, esp. from a simpler to a more complicated or advanced state; the passage of something through a succession of stages. Also, origination by natural development as opp. to production by a specific act.

This definition is radically different from its politically correct cousin. Evolution is no longer a theory open to scientific debate and the skeptic's doubt: it is now a *process* produced by the natural development rather than creative acts performed by supernatural, extraterrestrial, or lexicographic agencies. It now has stages rising, except for when they don't, from the simple to the complex.

The equivocation is readily apparent to those skeptical of consensus. In *Discerning Truth: Exposing Errors in Evolutionary Arguments*,[121] a primer on logic for Christian homeschoolers, author Jason Lisle uses the mixing of definitions of evolution between the actions of "change" and "common descent" in a single argument as an example of the logical fallacy of equivocation, a sort of bait used for deception or persuasion. It would be equivocation, for instance, to declare the *fact* of evolution as change is the consensus of the scientific community and then redefine evolution as organismic capitalism.

Natural Selection

Thanks to Darwin, evolution may be an accepted belief of the orthodox scientific community, but in Darwin's day, at least, natural selection was merely a proposed mechanism for evolution. In the words of Jerry Coyne:

> Although the idea of evolution itself was not original to Darwin, the copious evidence he mustered in its favor convinced most scientists and many educated readers that life had indeed changed over time. This took only about ten years after *The Origin* was published in 1859. But for many years thereafter, scientists remained skeptical about Darwin's key innovation: the theory of natural selection.[122]

Lest schoolchildren lose confidence in our court system, Coyne encourages us with the strength of his conviction: "Today scientists have as much confidence in Darwinism as they do in the existence of atoms."[123]

Coyne is not alone in his faith in Darwin's organismic capitalism. Take evolutionary psychology advocate Steven Pinker. Pinker, a Harvard professor, loves natural selection. He is certain: "Natural selection has a special place in science because it alone explains what makes life special."[124] He hypothesizes, "Better vision leads to better reproduction."[125] And he heartily believes:

> Because there are no alternatives, we would almost have to accept natural selection as the explanation of life on this planet even if there were no evidence for it. Thankfully, the evidence is overwhelming. I don't just mean evidence that life evolved (which is way beyond reasonable doubt, creationists notwithstanding). But that it evolved by natural selection.[126]

The skeptic cannot help but have doubts. How can science, much less natural selection, predict what smile will make life special? What correlation is there between better vision and family size? Couldn't marrying an ugly man by mistake be a blessing? Finally, how can we help but doubt that a deep time capable of evolving marching bands from microbes could, in the deep future, not evolve a plausible alternative theory to natural selection? Still, the skeptic can't help but admire Pinker's use of enthusiastic certainty to carry his rhetoric.

Two Stories, One Title

Confusing the word *evolution* meaning "change" with the word *evolution* meaning "natural selection" is sound equivocation, especially since the term *natural selection* is itself ripe for further equivocation. Evolutionist Hugo de Vries (1848-1935) captured part of the term's versatility his 1904 *Species and Variation: Their Origin by Mutation*: "Natural selection may explain the survival of the fittest, but it cannot explain the arrival of the fittest." In the arrival story, natural selection built humans and cockroaches one tiny block at a time from a common ancestor using organismic capitalism to discard all but the best blocks. In the survival story, if a thermonuclear extinction event favored the cockroaches, they, not the humans, would be deemed "the fittest."

Pinker favors the first version of the natural selection story, the organismic capitalism story:

> Natural selection remains the only theory that explains how adaptive complexity, not just any old complexity, can arise, because it is the only non-miraculous, forward-direction theory in which *how well something works plays a causal role in how well it came to be* [emphasis added].[127]

The skeptic can easily agree that nothing about the theory, outside of its acceptance, is miraculous; believing the "how well something works" part of the story comes harder. The assumption works better for some versions of natural selection than others.

The second version of the natural selection tale was championed by English philosopher Herbert Spencer (1820-1903). He coined the phrase "survival of the fittest." Darwin had better sense than Spencer and clung to natural selection for five editions of *Origin* before succumbing to the suggestions. Following consensus is never without it hazards. Survival of the fittest may be fine for ballet or shot put; the phrase may even work for Shakespeare's plays, which continue to survive to this day. Survival of the fittest, however, runs into real problems when Darwinian natural selection is combined with Mendelian genetics, a combination referred to as "the modern synthesis."

Unfit Fitness

In the synthesis, the definition of survival changes, allowing for easier equivocation. According to George Gaylord Simpson (1902-1984), one of the most influential paleontologists of his century:

> Natural selection favors fitness only if you define fitness as leaving more descendants. In fact geneticists do define it that way, which may be confusing to others. To a geneticist fitness has nothing to do with health, strength, good looks, or anything but effectiveness in breeding.[128]

Despite Pinker's enthusiasm for "how well something works plays a causal role in how well it came to be," in genetics fitness is not defined by intelligence, aptitude, creativity, kindness, or contribution to the world, nothing to do with Darwin's thought on structural advantages in the struggle for survival. Fitness is quite simply reproductive success. Any genes Shakespeare failed to share with his sister had no destiny, for all his descendants died. His sister's descendants survive even today. In the eyes of geneticists, Shakespeare was a misfit, his sister a veritable genetic Miss Universe.

The Department of Redundancy Department

If fitness is viewed as survival through offspring, then survival of the fittest becomes survival of the survivors. If survival is viewed aₒ evolutionary success and fitness is viewed as procreative success, then survival of the fittest becomes success of the successful.

To explain the skeptic's dissatisfaction with such phrases, a trip to the department of redundancy department is in order. When it comes to repeating ourselves, the word *tautology* is used to describe the act of saying the same thing again and more than once. *Tautology* comes from the Greek *tautologia*, from *tauto* meaning "the selfsame" and *logos* meaning "word." Consider these statements: "Those who succeed are successful." "Those who survive are survivors." "Immortals are not mortal." Such declarative statements depend on tautology. By definition, immortals eschew mortality, the survivors have survived, and the successful have succeeded.

Who can argue?

The claim that survival of the fittest is a tautology is nothing new; it is a favorite of Darwin's critics. Former University of California at Berkeley law professor and father of the intelligent design movement Phillip Johnson, in his book *Darwin on Trial*, presents a well-argued prosecution of Darwin's theory, quoting— besides George Gaylord Simpson above—any number of famous evolutionists as expert witnesses to testify that the survival of the fittest is a tautology.

In fairness, Elliot Sober, Hans Reichenbach Professor of Philosophy at the University of Wisconsin at Madison, dismisses the charges, brilliantly arguing that in philosophy, at least, tautologies are expressed by declarative sentences (*P or not-P*), "but notice that the phrase 'survival of the fittest' is not a declarative sentence."[129] Hans Reichenbach would have been proud, but, just in case, Sober goes on to strengthen his case, arguing that mathematics is tautology (always repeating itself after the equal sign), and concludes, "There is no point in withholding the label of 'science' from evolutionary biology just because it isn't exactly physics. Of course the theory contains 'tautologies' (mathematical truths); every theory does."[130]

While not even a skeptic could doubt that evolutionary biology is not rocket science, we certainly need to question my summary of Sober's arguments. If in doubt, the skeptic should read the philosopher personally, as I did after coming across his name in a Gould quote:

> This hoary claim ['the old canard that natural selection is a tautology and therefore empty of content'], still a favored gambit of creationists, brands selection as a useless concept because its watchword—'survival of the fittest'—becomes meaningless when fitness is defined in terms of survival. The argument can

be refuted in several ways (including the value of tautology in many scientific contexts—see Sober, 1993), but Darwin's own rebuttal seems most compelling to me. Darwin did not define fitness retrospectively by observed survival. He insisted, in principle at least, that fitter organisms could be identified before any environmental test by features of presumed biomechanical advantage. (The speediest deer can be specified beforehand, and their differential survival in a world of wolves can then be tested empirically.)[131]

For whatever reason, Gould did not quote Sober directly, but more importantly Gould tells us that Darwin was not interested in survival of the fittest in which fitness is defined as survival. Darwin was interested in arrival of the fittest—the building of new organisms, one stronger beak, one longer wing, one more successful biomechanical variation at a time. Confusing these two very different uses of the term *natural selection* has become another equivocation on the order of confusing evolution as change over time with evolution as microbes to marching bands, or confusing evolution with natural selection. Survival of the fittest is a tautology. Arrival of the fittest is not.

Making things even more complicated, it seems there are, again, two principle competing stories used to define arrival of the fittest. Darwin and his disciples—including Dawkins, Dennett, Coyne, and Pinker—favor one arrival of the fittest story. Gould includes the other story as an important component of the possible truth. The skeptic is wise to examine both of these stories, lest the phrase itself be used for the purposes of equivocation.

Adaptationism

Darwin was an adaptationist; he saw that organisms were adapted to their environments. Darwin was fortunate in that the pieces can always be made to fit when predicting the past. Bars are full of drunks, not teetotalers, because once upon a deep time drunks were adapted to hold shot glasses and teetotalers, teacups. Natural selection is always presumed to be the molding force. The anecdotes all sound the same: the fastest deer, the sharpest beak, the advantage of spur or horn, and the correctly-shaded moth all live to have more and better sex to pass on their advantageous variations to offspring, who will also have more and better sex ("better" being defined as "producing more great-great-great-grandchildren").

Critics of adaptationism have been with us ever since Darwin's theory first appeared. Those who would like to promote the myth that natural selection is the backbone of genetics, seldom mention that William Bateson (1861-1926), who invented the word *genetics* and founded the *Journal of Genetics*,[132] was no fan of the "astonishing ingenuity" of Darwinian "storytelling."[133] Over a hundred years ago, Bateson questioned Darwinian adaptationist tales:

> In any case of variation there are a hundred ways in which it may be beneficial or detrimental. For instance, if the 'hairy' variety of the moorhen

became established on an island … I do not doubt that ingenious persons would invite us to see how the hairiness fitted the bird in some special way for life in that island in particular. Their contention would be hard to deny, for on this class of speculation the only limitations are those of the ingenuity of the author.[134]

As any good skeptic should realize, these explanations are especially suspect when they are applied to the mind, which leaves no fossil record at all. No one does a better job at telling these stories than our expert on evolutionary psychology, Steven Pinker:

> Given that our brains were shaped by natural selection, it could hardly be otherwise. Natural selection is driven by the competition among genes to be represented in the next generation. Reproduction leads to geometric increase in descendants, and on a finite planet not every organism alive in one generation can have descendants several generations hence. Therefore organisms reproduce, to some extent, at one another's expense. If one organism eats a fish, that fish is no longer available to be eaten by another individual.[135]

Despite the flourishing of the neuronal brand of social Darwinism, Pinker continues to lament, "People desperately want Darwinism to be wrong."[136]

The skeptic wonders. If evolutionary psychology rests on the hypothesis that our brains were shaped by natural selection, might evolutionary psychologists, whose worldviews and livelihoods therefore depend on natural selection, desperately want Darwinism to be *right?* Just how desperate some are is captured in *What*

Darwin Got Wrong when the book discusses what is accepted and rejected by today's scientific community:

> Entire departments, journal and research centres now work on this principle [rejecting contradictions to Darwinism]. In consequence, social Darwinism thrives, as do ... psychological Darwinism, evolutionary ethics—and even, heaven help us, evolutionary aesthetics ... [W]hat we think that is needed is to cut the tree at its roots; to show that Darwin's theory of natural selection is fatally flawed.[137]

A noble goal. The skeptic is always eager to cut certainty at the roots.

Inner Strength—Structuralism

Among evolutionists, there is a second group, the structuralists.[138] Adaptationists such as Darwin looked outward, reasoning that a personal natural selection chiseled structures to fit the environment. Structuralists look inward, enjoying a different story, a non-Darwinian tale. They believe that organisms have been molded internally by their structures. Development being what it is, only mutations that get along with an organism's internal structure ever see the light of day. The primary competition is within, not without.

Darwinian adaptationists tell a story in which variations found among organisms occur by chance, in no particular direction ("isotropy of variation"). There is a problem with this. When Richard Dawkins argues how an aborted fetus "surely suffers less than, say, an adult cow in a slaughterhouse,"[139] he reassures us that "the majority of conceived embryos spontaneously abort anyway. It is probably best seen as a kind of natural 'quality control.'"[140]

So here's the question: how could Darwin's biomechanical variations be random if before entering the world they had already undergone quality control? Dawkins has mutated Darwin's story in a very necessary way if structuralism is to be avoided. For Darwin, competition occurred mainly at the level of the organism. For Dawkins (and Pinker quoting Dawkins' mutation of natural selection), competition begins at the level of "selfish genes"—personified chemical chains born with a drive to succeed—or so the metaphor goes. Dawkins substitutes a personal gene for personal natural selection or a personal God—depending on how personal the skeptic wants to get.

What the skeptic needs to understand is that for the purposes of selfish gene debates, life begins at conception. Thus, the structure of the body can become the selfish gene's environment, and the internal can become the external. Rhetoric has its advantages, but again, to replace arrival of the fittest by organismic capitalism with arrival of the fittest by genetic capitalism is just another facet of evolutionary equivocation.

Hopeful Monsters

Perhaps the most gifted thinking-outside-of-the-box evolutionist of the twentieth century, Stephen Jay Gould devotes nearly 500 pages to a historical treatment of the arguments between evolutionists who supported structuralism versus adaptationism. Gould devotes an episode to the tragic tale of Richard Goldschmidt (1878-1958), whose 1940 book *The Material Basis of Evolution* proposed that a simple but extremely rare alteration in a gene controlling the rate of development could produce a profound change. This change would allow a "hopeful monster" to emerge—a monster leaping ahead of natural selection's tiny steps, a monster with hopes of originating a new species, if not a new branch on the tree of life. How this monster would have gone about getting a date is a *Beauty and the Beast* tale deserving Disney animation. Overlooking the potential, the Darwinian adaptationists ridiculed Goldschmidt. As Gould recalls from his days as a graduate student at Columbia University, beginning in 1963:

> I had never heard of Richard Goldschmidt. Yet his name surfaced in almost every course—never with any explication of his views, but only in a fleeting and derisive reference to something called a 'hopeful monster.' Students then responded with a derisive sign of recognition—as our professors seemed to expect as a badge of membership in some inner circle.[141]

Goldschmidt hypothesized that species were separated by "bridgeless gaps," that natural selection might cause variation within a species, but it bore "no causal relevance to the production of new species."[142] Should Goldschmidt gain acceptance, Darwin's version of natural selection creating the arrival of the fittest—the variations between the major branches in the tree of life, the differences, say, between vertebrates and invertebrates or starfish and star gazers—would be recognized as a myth. Not only does Goldschmidt's argument strike at the heart of adaptationism; it strikes at another precept that Darwin demanded, gradualism.

No Leaps of Faith or Form

Darwin's natural selection insisted that the changes in the forms of the organisms must be Gradual with a capital G: "We see nothing of these slow changes in progress, until the hand of time has marked the long lapse of ages and ... the forms of life are now different from what they were."[143] Baby steps are the price we pay for the comfort of Darwin's natural selection sculpting life's diversity. That single longer giraffe vertebra was evidently invisible to all but the millimeter sticks of fawning future mates.

The word *saltation* is derived from the Latin *saltare*, meaning "to leap." A saltation is a sudden leap in form. Following tradition, Darwin quoted, *"Natura non facit saltum"* (Nature makes no leaps). He wrote, "for natural selection can act only by taking advantage of slight successive variations; she can never take a leap, but must advance by the shortest and slowest steps."[144] Darwin further insisted, "If it could be demonstrated that any complex organ existed, which could not possibly have been formed by numerous, successive, slight modifications, my theory would absolutely break down."[145] In other words, if it could be shown that the giraffe's neck sprang heavenward in a single developmental leap rather than a single slightly longer vertebra at a time, then Darwin's theory could be added to the tales of the Brothers Grimm. Adding the leap of neck to the leap of tautology—hypothesizing that the chance fit of the sudden elevation of view succeeded because it succeeded or survived because it survived—would still defeat Darwin's theory and still qualify as equivocation. The neck of Darwin's giraffe would join the neck of Lamarck's giraffe in the stocks reserved for biological fantasies and ridicule.

To say hello to saltation is to say goodbye to Darwinism—at least any Darwinism imagined by Darwin or his more devout disciples. Chance events—so extremely rare as to almost qualify as miracles—would once again be doing the sculpting, not gradualism. In the words of Professor Jerry Coyne, "Now when we say that 'evolution is true,' what we mean is that the major tenets of Darwinism have been verified. Organisms evolved, they did so gradually ... No serious biologist doubts these propositions.[146] The statement is undoubtedly true, not only because the court system agrees, but because Gould was a paleontologist, not a biologist. His claims that the fossil record does not stack up gradually as Darwin imagined, but in fits and starts; the theory is called "punctuated equilibrium." Replacing gradualism with punctuated equilibrium and continuing to call it Darwinism is, again, evolutionary equivocation.

A Monstrous Attack of a Theory

Rather than pursue Gould's tale, we will consider a thought experiment. An evolutionary paleontologist comes across Siamese twins in the fossil record. If a saltationist, the scientist would calmly assume the conjoined appendage developed in a single leap. If a Darwinian, however, the scientist would have to produce the missing links needed to maintain the dogma of gradualism—fossils with extra toes and fingernails, fossils with extra hands, fossils with an extra torso, etc.—not that this would present any problem. The Siamese twins could always serve as a missing link somewhere else. In the fossil record, how can the skeptic tell if nature is equivocating—selling isolated errors of development for actual genetic mutations? How can the skeptic know we are not confusing Siamese twins as steps in the ladder of evolution? Those producing rare examples of "missing links" seldom bother to ask such questions, and apparently hope that the skeptic won't either.

Did evolution come from within or without organisms? The consequence of the debate is best summarized by another quote from *What Darwin Got Wrong*:

> This view [gradualism] is seriously defective if, as we suppose, the putative random variations are in fact highly constrained by the internal structure of the evolving organisms. Perhaps it goes without saying that the more the internalist story is true, the less work is left for appeals to natural selection to do.[147]

In other words, confusing evolution with natural selection may well be the equivocation of equivocations.

Postmodern Evolution

What separates biomythology from real science? Real science inspires engineers to control the natural world; biomythology allows biologists to engineer the beliefs of the public. An excellent example of such biomythological antics appeared in *Nature* under the title of "Postmodern evolution?"[148] The article chronicles an Altenberg, Austria meeting of high-profile researchers discussing the future course of evolutionary theory. It begins with evolutionist Massimo Pigliucci, cautioning his fellow evolutionists, "If there's one thing we don't want, it's for people to get the idea that there's a bunch of evolutionary theories out there and they are all equal." Pigliucci obviously isn't worried about scientists accepting such relativism; the people he is referring to are the public. The battle is not about the internal progress of scientific knowledge; it's about the rhetoric of selling Darwin to the greater culture. Such a concern is biomythology, pure and simple.

The article's concluding paragraph stresses the continuing paranoia about damage to the image of "the modern synthesis."[149] Jerry Coyne cautions, "People shouldn't suppress their differences to placate the creationists, but to suggest that neo-Darwinism has reached some kind of crisis point plays into creationists' hands."[150]

David Cook

In other words, don't confuse the public with the doubt and uncertainty inherent in the scientific method.

In summary, Spencer's tautological survival of the fittest is true wherever winners win, persuaders persuade, and the successful have succeeded. Darwin's version of arrival of the fittest may one day be remembered as the founding myth of a historical movement. If the skeptic listens carefully whenever a rhetorician mixes the two—claiming that a lone mutation sent the giraffe's neck soaring upward in a single leap, a forbidden saltation, and *then* the survivors survived—maybe he or she can hear Darwin turning over in his grave.

The Skeptic's Guide

Every skeptic should watch for evolutionary equivocation. When promoting Darwinism, it is equivocation to offer change over time and then deliver the microbes *to* marching bands of common descent with modifications. It is equivocation to sell evolution and deliver natural selection. It is equivocation to sell arrival of the fittest and deliver the tautological survival of the fittest. It is equivocation to sell adaptationism and deliver structuralism. It is equivocation to sell Darwin's gradualism and deliver Gould's punctuated equilibrium. When the skeptic surveys history, the only unequivocal definition for evolution may well turn out to be "change in equivocation over time."

Chapter 6
Murder by Ugly Fact

"No single theory ever agrees with all the facts in its domain."
—Paul Feyerabend, *Against Method*

No primer on skepticism would be complete without paying a visit to the words of T. H. Huxley (1825-1895). Huxley was known as "Darwin's bulldog" because of his loyal defense of Darwin and evolution, if not natural selection. As we mentioned in our first chapter, Huxley also coined the term *agnostic* for those who can't really know, who refuse to believe until they reach purgatory and see the wounds for themselves. Whether Huxley's sympathy for evolution and agnosticism sprang from a common materialistic passion against those who were too eager to preheat the brimstone for the benefit of heretics, I can't say. He was, however, the author of one of science's most endearing myths. In "Biogenesis and Abiogenesis," from his 1870 collected essays, Huxley wrote, "The great tragedy of science—the slaying of a beautiful hypothesis by an ugly fact."

However fond Huxley's admirers are of the quote, can an ugly fact slay a beautiful hypothesis? The myth is too precious! It is repeated as a rhetorical flourish whenever one wishes to unbalance skepticism by increasing doubt about religion, but not science. The faithful, as the myth goes, are blinded to facts, however beautiful, by their ugly dogmas. Scientists are blinded to theories, however beautiful, by ugly facts. Evidently, the difference between a dogma and a theory is how they fit into the ugly fact myth.

Ain't No Martian Bones in the Precambrian

Richard Dawkins states the myth succinctly:

At any moment somebody might dig up a mammal in the Cambrian rocks [a geological period in which the theory of evolution tells us no mammals yet existed], and the theory of evolution would be instantly blown apart if they did. [151]

Jerry Coyne discovered the same myth:

Despite innumerable possible observations that could prove evolution untrue, we don't have a single one. We don't find mammals in Precambrian rocks, humans in the same layers as dinosaurs, or any other fossils out of evolutionary order. [152]

New Yorker journalist Adam Gopnik shares the priority for the same discovery:

The theory of evolution by natural selection is an argument: all its points are open, its claims clear, many of its possible refutations self-evident. (Find the fossilized body of a Pekingese lapdog in the Pleistocene, and we all start over.) [153]

Finally, skeptic supreme Michael Shermer fleshes out the idea and even provides an auxiliary hypothesis, just in case an anomaly[154] occurs:

> If such a fossil juxtaposition occurred [a fossil horse and trilobite in the same stratum] and it was not the product of some geological anomaly (such as uplifted, broken, bent, or even flipped strata—all of which occur but are traceable), it would mean that there was something seriously wrong with the theory of evolution.[155]

Is the astonishing agreement between the four authors' examples evidence of their descent from a common Precambrian ancestor? As with evidence in general, other interpretations are possible.

Karl Popper Leaps Tall Science with a Single Bound

What marks the boundaries of science? No exact agreement exists among philosophers, but the Darwinians apparently love twentieth-century philosopher Karl Popper's take on the subject. Both Jerry Coyne in *Why Evolution is True*, 2009, and Richard Dawkins in *The Greatest Show on Earth: The Evidence for Evolution*, 2009, allow Popper as an expert witness in the trial to determine what constitutes real science as opposed to religion.

Popper—knowing full well that the partial evidence available to us proves nothing absolutely—suggested that while a scientific theory cannot be proved, it must nevertheless be capable of falsification; that is, it must be falsifiable. By this, he meant that one could design an experiment that could conceivably disprove or "falsify" the theory. If, for instance, your theory was that all swans are white, your experiment might be to search Central Park for purple swans. A success would falsify the white swan theory. A failure would not rule out purple swans preening in Mexico or black swans swarming in Australia. The experiment would prove nothing universal, but merely fail to falsify the theory in the setting at hand. According to Popper, such experiments need not be run. They need only be describable. Popper developed this boundary to separate metaphysics from empirical (observable) science. After Popper, the art of science became arranging observations so as not to falsify theory. The art works well.

So how do we falsify evolution? J. B. S. Haldane, a revered Darwinian of the twentieth century, "famously retorted, when asked to name an observation that would disprove the theory of evolution, 'Fossil rabbits in the Precambrian!'"[156] In the Darwinian worldview, the head bone was not connected to the backbone before the Cambrian period (about 500 million years ago). Natural selection or aliens had not yet invented backbones. At that point in history, our ancestors were all spineless. Peter Cottontail hopping down the Precambrian bunny trail would therefore falsify the theory of evolution.

Darwinians brandish the Popper-Haldane story to discard intelligent design, the popular theory that rather than being self-designed the universe was designed by a designer, an alien rabbit perhaps, that long ago hopped out of the Precambrian. Since you cannot falsify such a designer, intelligent design, we are told, cannot be science. Does the old rabbit-in-the-Precambrian ploy, however, really insulate Darwin from the beauty of the Ugly Fact Myth?

Deep Time is Not God; the Fossil Record is Imperfect

Our Siamese twin thought experiment in the last chapter might provide a challenge for a lessor Darwinian, but Darwin himself had another trick up a sleeve of his theory that could be used to explain why the body parts of the Siamese twin did not appear gradually in the fossil record. He had the "extremely imperfect geological record":

> All these causes [incomplete exploration, unpreserved classes, fossils not made during periods of maximum variation, movement of organisms] taken conjointly must have tended to make the geological record extremely imperfect, and will to a large extent explain why we do not find interminable varieties, connecting together all the extinct and existing forms of life by the finest graduated steps.[157]
>
> He who rejects these views on the nature of the geological record, will rightly reject my whole theory.

Thus, to reject the auxiliary hypothesis that the fossil record is extremely imperfect is also to reject Darwin's theory.

There is a problem that the Popper-Haldane enthusiasts seldom mention when demoting intelligent design from falsifiable science. Popper also stipulated, "As regards *auxiliary hypotheses* we propose to lay down the rule that only those are acceptable whose introduction does not diminish the degree of falsifiability of the system in question."[158] For Darwin, natural selection was a hypothesis to explain the changes over time known as evolution. Verifying natural selection in the fossil record relies on an auxiliary hypothesis: the geological record is extremely imperfect.

So how do we falsify that the geological record is extremely imperfect? What experiment could be run to disprove that the geological record is imperfect? The old "black swan in the Precambrian" ploy doesn't work. Yet to reject this auxiliary hypothesis as science is again—by Darwin's insistence—to reject his whole theory.

Instead of placing his faith in God—who cannot be falsified—Darwin placed his faith in the extreme imperfection of the geological record, which also cannot be falsified. Like God, the non-falsifiable auxiliary hypothesis explains away all Darwin's trials and tribulations. Why do the gradual tiny steps between species characteristically leave no trace? The geological record is extremely imperfect. Why hasn't an ugly fact killed Darwin's beautiful hypothesis? The ugly fact was lost in the geological record because *the geological record is extremely imperfect!* It washes away all the sins of paleontology.

If we are going to cite Popper as an expert witness, who then in science are the true Popperians? Who devote themselves to keeping skepticism balanced, to falsifying rather than promoting Darwin's theory? Philosopher Thomas Nagel provides the answer:

Nevertheless, I believe the defenders of intelligent design deserve our gratitude for challenging a scientific world view that owes some of the passion displayed by its adherents precisely to the fact that it is thought to liberate us from religion.[159]

Thus, if science must be falsifiable, and the intelligent design movement revolves around falsifying Darwin's theory, and the apologists for Darwin devote themselves to using the universal solvent of rhetorical reason to dissolve away such falsifications, who, according to Popper, would be the real scientists?

Darwinian Myth Expansion: Falsify That!

Darwin provided an entire chapter, "Difficulties on Theory," in which he used the universal solvent to dissolve away obvious difficulties with his primary myth. Darwin's disciples similarly share a knack for employing what I will call *myth expanders*.

If natural selection favors variations, why are some variations about as useful as teats on a boar? Because the variations are, like teats on males, along for the thrill of the ride.

Falsify that!

Why do organisms have variations that are not favorable to the environment at hand? Because the variations were favorable to an earlier environment (like heaven) of which we have no real knowledge.

Falsify that!

If organs were developed one tiny step at a time, what good would a tenth of a wing be for flying? The tenth of a wing (like a single longer giraffe vertebra) was good for something else—scratching a girlfriend's back, perhaps? Or how about Steven Pinker preserving the Darwinian myth of biomechanical gradualism? "Perhaps the incipient wings of insects first evolved as adjustable solar panels which soak up the sun's energy when it is colder out and dissipate heat when it's warmer."[160]

How do we falsify this story? Did Pinker go back to when wings first evolved to interview a few insects and see if they yearned for sweaters or fans?

Finally, how did natural selection justify the cost of dental care for saber-toothed tigers? Female tigers found them sexy. (How do you falsify what was on the minds of female tigers 20 or 30 million years ago? I can't be sure what was on my wife's mind last night at dinner, much less falsify it.)

Spandrels

Stephen Jay Gould provided another tool of myth expansion. In addition to suggesting that Darwin's conception of gradualism was wrong—that the incompleteness of the fossil record was not the same as a failure to represent the march of evolution, and that long periods of equilibrium were actually punctuated with short periods of change—Gould gave us the spandrel.

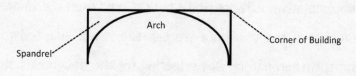

Figure 1

In architecture, a spandrel is a sort of figure-ground phenomenon. In Figure 1, either the arch is the figure and the spandrels the ground, or the spandrels can be the figure and the arch be the ground. Thus, Gould asks if Mother Nature is selecting for the figure or the ground. Whenever the figure cannot be explained by adaptationism, then we can imagine that the figure was not the figure but some unimagined ground of the figure that was adapted.

For example, it turns out the same OtxI master gene controls for both the development of the external genitalia and the thickness of the cerebral cortex[161]—the little head and the big head. That presidents such as Roosevelt, Eisenhower, Kennedy, Nixon, and Clinton sometimes thought with the wrong head is justified by evolutionary psychology: It wasn't a mistake. It wasn't a sin. It was an evolutionary spandrel.

The spandrel idea is hardly original with Gould. It dates back to at least Saint Augustine (354-430 A.D.), who more or less argued that God created the figure of good; evil was merely the ground. God gave us the sacred non-material, non-fattening doughnut hole to contemplate. The evil, fattening doughnut was just along for the ride. If some of us chose evil over good and are now fat rather than contemplative, it is not God's fault. Gould used the same idea for evolution: if the doughnut we can see is not adapted to the environment, then nature was selecting for the doughnut hole we cannot see.

Falsify that.

The point here is that whenever a Darwinian or evolutionary psychologist imagines a scenario of how something went down in the past, the skeptic should decide if a falsifiable hypothesis is being offered. Then the skeptic should accordingly place the scenario in the rectangular file of real science or the circular file of biomythology. Each time a tale has to be propped up by another, the well-trained skeptic should similarly ask, "How would you falsify that auxiliary hypothesis?" If falsifying the anecdote is more difficult than falsifying the theory itself, then toss the anecdote; it's rhetoric, merely metaphor, no more a part of universal science than are the gods of Mount Olympus.

Immunity from Ugly Facts

Popper was all too aware that scientific inoculations might allow an ugly fact to falsify a hypothesis but still not kill it. In summarizing the arguments of his opponents who denied that a scientific system could, in reality, be falsified, he wrote:

> Thus we may add *ad hoc*[162] hypotheses. Or we may modify ... [the definitions]. Or we may adopt a skeptical attitude as to the reliability of the

experimenter whose observations, which threaten our system, we may exclude from science on the ground that they are insufficiently supported, unscientific, or not objective, or even on the ground that the experimenter was a liar ... In the last resort we can always cast doubt on the acumen of the theoretician.[163]

We change our myths as we go, creating *ad hoc* arguments to dissolve away—with the universal solvent of reason—embarrassments to theory. As Popper suggests, we can deny anything I say about biomythology by simply denying my credentials. Also, we can, as many have, deny Popper's argument as valid. Anyone, such as Dawkins or Coyne or Gopnik or Shermer, who uses falsification as an argument for kicking intelligent design out of science inadvertently admits Popper as an expert witness. Not all Darwinian apologists are so foolhardy.

As mentioned, at the end of the first decade of the twenty-first century, three books offering Darwinian apologias were published. Coyne and Dawkins use the Popperian defense. In the wisest and most thoughtful of the three, *Only a Theory: Evolution and the Battle for America's Soul*, author and Brown University biology professor Kenneth R. Miller showed better sense. He did all he could to distance himself from Popper and diminish the philosopher's reputation:

> Cognitive relativism is essentially an argument that truth and falsehood are not the absolutes we often take them to be, but rather are constructed by individuals and social groups. Karl Popper, the great

philosopher of science, had helped to pave the way for this relativistic foray into science by challenging its cherished notions of objectivity ... If objective truth is a myth, then what does this make of science in general? The answer should be perfectly clear: not much.[164]

As my writing makes clear, I always love destroying a reputation with a good straw-man argument. Still, it is doubtful that Popper felt the rules of aerodynamics were culturally dependent, fine for New York but not Alabama. Miller is a realist, not an instrumentalist. It's not enough that a theory inspires a workable technology; we must *believe* that the theory is *true*—whether or not is has inspired any technology. When it comes to realists, skepticism is unwelcome whether your god is God or the scientific method.

For example, philosophy professor James Hall writes:

Reality is not in the eye of the beholder. It may be filtered through the eye of the beholder, but it retains an intractable content. Hard heavy objects break bones and cause pain, think what we will.[165]

Fortunately, mental objects hurt less than physical objects. Stories are not to be confused with actions. Actions are clear; which story is being acted out is not. Suppose that a marksman aims at a child and kills, by mistake, the bear poised to maul and devour the child. Now suppose instead that the marksman aimed at the attacking bear and killed the child by mistake. The fact of the slain bear coupled with the story about heroic marksmanship beats the fact of slain child coupled with the story about incompetent marksmanship every time. Baring eternity—actions are forever; the stories about actions change and *may*, when necessary, warrant our skepticism.

In the case of Professor Hall's defense of realism, bricks inflict the same amount of pain whether the story is that they are solid or primarily constructed by space between atomic particles. It's not the truth of the story that causes the pain; it's the truth of the action, the brick parting the sea of space. That is what's real. Only the most extreme skeptic doubts all actions beyond his or her private telling of stories. Most skeptics doubt stories, not actions seen with their own eyes. If a boy is stoned to death with a flying brick, he is just as dead if the story is execution, murder, gang war, or shot-put accident. Stories change; actions persist.

Whatever the truth of my story placing actions above stories, Dawkins and Coyne created a problem for themselves when they admitted Popper as an expert witness. To do so was to throw out "the geological record is extremely imperfect." As previously quoted, Darwin wrote, "He who rejects these views on the nature of the geological record will rightly reject my whole theory." Thus, with Darwin's full permission, if we reject the extremely imperfect geological record, the skeptic can reject the whole theory. Luckily for the Darwinians, that the imperfection of the fossil record cannot be falsified has not gotten it expelled from science. Richard Dawkins still loves the auxiliary hypothesis. He continues to write, "Only a tiny fraction of corpses fossilize, and we are lucky to have as many intermediate fossils as we do."[166] It is apparently fine if an ugly theory kills the skepticism of a high school student, so long as it does not kill Darwin's beautiful theory.

Further Ugly Facts about Ugly Facts

Besides the problems of aliens hiding in the imperfect geological record of the Precambrian, Bacon, almost four hundred years ago, had predicted another problem with the Ugly Fact Myth:

> Once a man's understanding has settled on something (either because it is an accepted belief or because it pleases him), it draws everything else also to support and agree with it. And if it encounters a larger number of more powerful countervailing examples, it either fails to notice them, or disregards them, or makes fine distinctions to dismiss and reject them, and all this with much dangerous prejudice, to preserve the authority of its first conceptions.[167]

Despite the advances of science, some things never change. Bacon's viewpoint of the stubborn resilience of beautiful theories was repeated by Darwin:

> Although I am fully convinced of the truth of the views given in this volume ... I by no means expect to convince experienced naturalists whose minds are stocked with a multitude of facts all viewed, during a long course of years, from a point of view directly opposite to mine ... [But] I look with confidence to the future, to young and rising naturalists, who will be able to view both sides of the question with impartiality.[168]

Darwin knew his facts, however imaginative or ugly, would never erase the Creator from the minds of his contemporaries. Max

Planck (1858-1947) was another scientist who found out the hard way what Bacon and Darwin had long before surmised. Classical physics (the physics of Newton and his followers) did not predict the glow of hot metals at the ultraviolet[169] end of the spectrum. The problem was referred to as the "ultraviolet catastrophe."[170] Planck solved the problem by proposing that light energy was not continuous, but rather came in packets. Thus, you could have one packet of energy or two packets of energy, but not one and a half or two and a half packets of energy. Planck called the packets "quanta." In proposing his "quantum theory," Planck stepped on the toes of his classical physicist contemporaries. It hardly mattered that the ugly facts of the newer data did not fit the older theory. In his *Scientific Biography*, Planck later wrote of his experience that a "new scientific truth does not triumph by convincing its opponents and making them see the light, but rather because its opponents eventually die, and a new generation grows up that is familiar with it."[171] Thus theories, like life, depend on death to be complete.

Ugly Facts and Beautiful Theories Are In the Eyes of the Beholder

Historian of science, Thomas Kuhn wrote *The Structure of Scientific Revolutions*, which we will return to in greater depth later. Kuhn described what he called "paradigms"—essentially, the major agreements guiding science during a particular period of history. Newton's absolute time and space was one such paradigm. Einstein's relative time and space was a different paradigm during a different period of history.

Kuhn made a number of observations, all of which he supported with examples, about how ugly facts are handled. First, scientists do not reject theories based solely on ugly facts. For instance, "To be accepted as a paradigm, a theory must seem better than its competitors, but it need not, and in fact never does, explain all the facts with which it can be confronted."[172] Thus, there are always ugly facts, and theories nevertheless persist. Kuhn further notes, this time about Popper's falsification demarcation for science:

> But falsification, though it surely occurs, does not happen with, or simply because of, the emergence of an anomaly or falsifying instance. Instead, it is a subsequent and separate process that might equally well be called verification since it consists in the triumph of a new paradigm over the old one.[173]

Darwinian Fundamentalism

Can an ugly fact kill a beautiful theory? Not everyone would agree with Kuhn, but if a fossil mammal were dug up in the Precambrian rocks, the Darwinians would accept the discovery about as readily as the average Christian would accept the testimony of someone claiming to be the returned Christ. It would take far more than a single scarred palm or fossil. The claim would be fiercely contested and labeled as a hoax. The discoverer would be hotly discredited, his credentials scrutinized, his honesty maligned. The Darwinians would demand the discovery be duplicated by an avowed Darwinian. A single missing link can do in a pinch—say, if you need a developmental anomaly to serve as a stepping stone between dinosaurs and birds—but a single unwanted link demolishing a cherished theory would never be tolerated, not at least until a new myth had been devised and voted in by a consensus. No one likes the sting of chaos, even if that chaos stimulates a new beginning.

Respect for the Law

Should we be upset that the incomplete geological record necessary for Darwin's theory couldn't be falsified? Logic rests on a handful of basic assumptions. One of these is the Law of Non-contradiction, which Aristotle in *Metaphysics* assumed was "the most certain of all" principles:

> It is, that the same attribute cannot at the same time belong and not belong to the same subject in the same respect; we must presuppose, in face of dialectical objections, any further qualifications which might be added.[174]

Does this prevent biomythologists from believing some ugly facts are beautiful? Not in the slightest. As any skeptic should now be able to tell you, the Law of Non-contradiction cannot be falsified and therefore, like God, has no place in science—which is why logic, as we will see, applies in science except for when it does not.

Chapter 7

Is Darwinism a Religion?

The Debate

Is water wet? Do crows fly? Ask. The inanity of the questions may raise an eyebrow or two, but the answers will evoke little emotion. Is Darwinism a religion? Ask. Now the skeptic will see temperatures, not eyebrows, rise. The responses will divide like the races on a 1950 Birmingham bus. At the back of the bus, evangelical Christians will inevitably nod their heads in agreement, smile, and say, "Of course it is. Darwinism is based on faith, just like Christianity. The difference is that while we own up to our faith, Darwinians are hypocrites, hiding behind the excuse of reason."

To the same question, the Darwinian, whether openly or privately exasperated, will affirm, with a little too much certainty, "Of course not. We live in light, you in superstition. Let's examine the facts." Then the Darwinians will recite a creed begun by Darwin in the "Natural Selection" chapter of *Origin's* original edition. In that chapter alone, Darwin repeats the words "I believe" no less than—count them for yourself—sixteen times. He also includes one "I do believe," one "I am strongly inclined to believe," one "I further believe," and one "we must believe," not to mention three "no reasons to doubt," one "I cannot doubt," and one "I do not doubt."

Why these writer's tics, this liturgy against skepticism? Was Darwin trying to convince his readers or himself? Was Darwin—yet unthawed by fame—still frozen midair in his leap of faith?

Will the future look back and see Darwinists as a zealous minority with reverence for the genius of their prophet? Will they be believed to have been re-educated away from the beliefs of their parents by years of "higher" indoctrination, distanced from society at large by a specialized language spoken only by the group, adamant that their own beliefs were true and good while the beliefs of the larger society were false and evil superstitions properly shunned by peer review?

Despite the similarities, for those who understand finance, Darwinism is definitely not a religion. As we explored in our third chapter, religion is funded by voluntary donations; Darwinism is funded by enforced taxation. The skeptic may refuse to finance Darwin, but only from behind bars.

So, if the skeptic fails to doubt the above argument and allow that Darwinism is not a religion, is it a cult? As Gopnick reminded us:[175]

> And this rootedness in reasoning explains why of all explanations of life, evolutionary theory is not remotely like a religion. There is no resemblance between evolutionary biology, even if we call it Darwinism, and religion. (And it is the devil's work to say it is!) The theory of evolution by natural selection ... deserves the respect we give to any wonderful and winning argument ... But it isn't a dogma, and the claim that it is made only by those

who want to protect their own faiths from criticism by pretending that all strong ideas can only take the form of faiths.[176]

Much of Gopnik's felicity I have already discussed. We spoke of real skeptics, who would undoubtedly doubt all "strong ideas" that are less strong than the noses on their faces. We considered how Darwin's "extremely imperfect geological record" is hardly open to self-evident refutation. If Gopnik demands the skeptic's respect for "wonderful and winning arguments," should we not extend this respect to all tales of transcendence, not just Darwin's deep time? Is the skeptic justified to suspect the wonderful-and-winning-argument argument is based on worldview rather than the etiquette of debate?

Dogma

As for Darwinism being or not being dogma, our good friends at Oxford define the term *dogma* as "Doctrines or opinions, esp. on religious matters, laid down authoritatively or assertively." Per this definition, is Gopnik's paragraph dogma *par excellence?* First, he is authoritative: not only is he an author, but he frequently appears in the *New Yorker*, a magazine qualifying as an authoritative, even enlightened, source by a segment of secular American urban culture. Next, he is not discussing Darwin so much as "religious matters." Finally, he is assertive: those who disagree with him, he reduces to "the devil's work." Despite his own dogmatic lapse, however,

Gopnick is hardly on firm ground when he equates his straw man religion with dogma, any more than I am when I skeptically equate my straw man biomythology with nonsense. When it comes to reliance on dogma, is today's Christianity any more encumbered than today's Darwinism?

Christian Dogma

Since the Enlightenment, respect for authority has dwindled. Today, some Christians handle wafers; some snakes. Some stress predestination; some free will. Some live in fear of hell; some in the joyful anticipation of eternity. Some see the Gospel story as our promise of salvation; some see the same story as proof of what happens when you put God in the hands of organized religion. Some judge acts; some judge people. Some reserve hell for those who disagree with them; some reserve hell for those who have harmed or demeaned others in God's name. Some think God is all; some think all is God. Some treasure icons; some are iconoclasts. Some seek an eye for an eye; some turn the other cheek. Some imitate; some adore. Some search the heart; some the history. Some prefer the book of nature; some the book of scripture. Some trust the spirit to interpret the Word; some trust the words of others to interpret the spirit. Some see God clearly reflected in scripture; some see Him reflected in a glass, darkly. Some believe in just wars; some settle for the sword of Jesus's tongue. Some lean toward works; some toward grace. Some believe that Jesus will have the final say; some that Jesus has already had his say. Some have arms weary from casting stones; some have arms strong from carrying and acknowledging their own sins. Some are intolerant; some are intolerant of intolerance. Some empty their minds with silent prayer; some fill their hearts with the Lord's Prayer. Some worship alone; some in groups. Some pray to the saints; some to the Lord. Some pray for the poor; some for prosperity. Some proclaim, "Faith alone, grace alone, Scripture alone"; some proclaim, "Love alone." In the words of nonbeliever Sam Harris, "People have been cherry-picking the Bible for millennia to justify their every impulse, moral and otherwise."[177]

So where exactly is the dogma?

Darwinian Dogma

Though it hasn't made it through centuries, much less millennia, the cherry-picking of Darwin is no less rampant. Stephen Jay Gould—whose ability to step back and view a larger perspective continually annoyed his more orthodox fellows—suggested that selective quotation to "support almost any position" is a habit of Darwinians as well:

> Since Darwin prevails as the patron saint of our profession, and since everyone wants such a preeminent authority on his side, a lamentable tradition has arisen for appropriating single Darwinian statements as defenses for particular views that either bear no relation to Darwin's own concerns, or that even confute the general tenor of his work.[178]

What Gould's words tell us is that the cherry-pickers of Darwinism are trying to make sense out of Darwin—or not—in light of today's observations, much as Christians may interpret Biblical passages differently with increasing experience, maturity, and reflection.

Just as philosopher of science Paul Feyerabend proposed a scientific method in which "anything goes," Darwin hardly constricted his subject more precisely, the final words of his introduction being, "Natural Selection has been the main but not the exclusive means of modification."[179] Christianity is diverse, and so is, with the full permission of its prophet, today's Darwinism.

In Jablonka and Lamb's *Evolution In Four Dimensions: Genetic, Epigenetic, Behavioral, and Symbolic Variation in the History of Life*, we find the caution:

> You get the impression that there is a tidy, well-established theory of evolution—Darwin's theory of natural selection—which all biologists accept and use in the same way. The reality is very different, of course. Ever since Darwin's book *On the Origin of the Species* appeared in 1859, scientists have been arguing about whether and how his theory of evolution works. Can competition between individuals with heritable differences in their ability to survive and reproduce lead to new features? Is natural selection the explanation of all evolutionary change? Where does all the hereditary variation on which Darwin's theory depends come from? Can a new species really be produced by natural selection?[180]

As we have seen, some Darwinians still believe in the organismic capitalism touted by Darwin and inspired by Malthus's grand plans for starving the poor. Some believe the competition occurs not at the level of the organism, but at the level of the gene, the family, the population, the ecosystem, or all of the above. Some believe that natural selection works its wonders on random variations like a longer vertebra of a giraffe's gradually lengthening neck; some believe that variations are not random at all but dictated by the internal structure of the organism. Some believe that the biomechanical variations of organisms change in tiny steps, as Darwin supposed (or at least this is the gospel told to high school students); some believe in *saltations*, that a single mutation may instigate profound leaps in timing and range of development and structural mechanics to produce "hopeful monsters." Some pay homage to the gradualism in geology that Darwin borrowed from his mentor Sir Charles Lyell[181] and demanded for his theory to be true; some believe in the fits, starts, and upheavals of Cuvier's[182] catastrophes—that a meteor, not competition between dinosaurs, jump-started what came next. Some believe in Darwin's arrival of the fittest; some rely on the tautology of the survival of the fittest. Some believe that the primary process of evolution is natural selection; some believe that natural selection, if it has any use at all, is only good for fine-tuning at the level of the species or, perhaps, the genus. Some believe that the fossil record is incomplete, or it would fully confirm Darwin's story. Some believe that the fossil record provides a representative sample and that Darwin's story needs to be modified. For some the persistence of foreskins despite centuries of circumcision rules out Lamarck's[183] inheritance of acquired characteristics even though Darwin thought enough of "use and disuse" to emphasize the concept by mentioning it in the final paragraph of his book; for some, Lamarck's inheritance of acquired characteristics might yet be supported by the science of epigenetics.[184] Some Darwinians believe that all life is connected; some believe that some lives are more connected than others are and that races are to be separated and pruned according to their relative "intelligence." Some believe that only the real Darwinians see the evidence as they do. Some believe natural selection means that because sustenance increases arithmetically $(2 + 2 + 2 + 2 \dots)$ and organisms increase geometrically $(2 \times 2 \times 2 \times 2 \dots)$ that organisms soon outnumber food, and those organisms with the best variations for their environments will win the competition with organisms with *bad* variations for their environments and will leave behind more offspring with good variations, and that this process has molded organisms to fit their environments and led to the construction of the entire tree of life. Some use natural selection to mean nothing more than "nature did it," or that a personal Mother Nature did it.

For two thousand years, people have been redefining Christianity to fit their own passions and purposes and times. For over one hundred and fifty years, Darwinians have been doing the same thing. To say that Christians or Darwinians are or are not dogmatic is rhetoric, not fact. Some are; some aren't. Who could be more dogmatic than New Age Christians banishing dogma? Even so, the followers of Jesus have not shared all dogma in common in almost a thousand years. Between the East-West Schism, the Reformation, and the New Age, any talk of universal dogma is an anachronism. To say otherwise is to compare medieval Christianity with modern biology. The comparison is about as fair as comparing bloodletting to contact lenses.

Darwinian Fundamentalism

Still, there are fundamentalists. Take Richard Dawkins' claim, "All reputable biologists go on to agree that natural selection is one of the most important driving forces [of evolution]."[185] Thus you are not a *reputable* biologist if you doubt Darwin's arrival of the fittest explaining the diversity of biological classification. Dawkins, of course, passionately argues that he is no fundamentalist:

> I am no more fundamentalist when I say evolution is true than when I say it is true that New Zealand is in the southern hemisphere. We believe in evolution because the evidence supports it, and we would abandon it overnight if new evidence arose to disprove it.[186]

Dawkins is apparently blind to the difference in reliability between a repeatable perception and an explanation. A tree falling in the woods may or may not make a sound without a mind. But, without a mind, an explanation cannot exist. Explanations are the stuff of idealism, not materialism. They are stories created by the mind for the mind—whether or not they conform to actions in the world. An explanation is hard-pressed to compete with a habitual action in a habitual environment. This is the reason that clinicians routinely ignore the results of evidence-based medicine that conflicts with their routine practices.

Perceptions are based on habit and so cannot be trusted beyond habitual circumstances—we are fooled by novel circumstances, by murders, magic, and optical illusions.[187] Nevertheless, a habitual perception confirmed by action in a habitual environment is more certain than an explanation waiting to be shot down by the next discovery. When it comes to certainty, the noses on our faces are one thing; natural selection constructing marching bands from microbes is quite another. With today's technology the direct, present-time perceptual knowledge of New Zealand's position on the globe is similarly at a higher confidence level than inference about an unseen evolution that will shortly be conflated with natural selection.

The Death of Purpose

Aristotle pictured a world in which we did not understand what we are looking at until we understood what it was made of, what moved it, what it formed, and what it was *for*. Everything had its purpose, its goal—an idea known as teleology. Teeth were created for chewing; noses, for picking. Aristotle knew that the alternative to purpose was necessity, the idea that the present was merely necessitated or determined by the past. Predating Darwin by a couple of millennia, Aristotle proposed (and rather quickly dismissed) the survival of the fittest:

> Why then should it not be ... that our teeth should come up of necessity—the front teeth sharp, fitted for tearing, the molars broad and useful for grinding down the food—since they did not arise for this end, but it was merely a coincident result; and so with all other parts in which we suppose that there is a purpose? Wherever then all the parts come about just what they would have been if they had come to be for an end, such things survived, being organized spontaneously in a fitting way; whereas those which grew otherwise perished and continue to perish."[188]

Thus, Aristotle hypothesized that, in a world without purpose, those variations that happened to fit into the chance organization of the system survived and those that didn't perished. Aristotle, passionately predisposed to purpose, dismissed the possibility, rationalizing that it was improbable that chance could have orchestrated it all. For many, alignment with purpose continues to define Truth. They are skeptical of detours.

Darwin saw things differently. He used evolution as a metaphor for a much larger purpose—the elimination of purpose. As historian of science Thomas Kuhn notes:

> All the well-known pre-Darwinian evolutionary theories ... had taken evolution to be a goal-directed process. The 'idea' of man ... was thought to have been present from the first creation of life, perhaps in the mind of God. That idea or plan had provided the direction and the guiding force to the entire evolutionary process. Each new stage of evolutionary

development was a more perfect realization of a plan that had been present from the start.

For many men the abolition of that teleological kind of evolution was the most significant and least palatable of Darwin's suggestions.[189]

Therefore, Darwinism is not just another variation on speculations about competing variations begun and dismissed by Aristotle two millennia earlier. Darwinism is the denial of a purpose that Aristotle believed informed life. Darwinism is a plea for the death of teleology,[190] at least a teleology based on theism or vitalism. Darwin's brand of evolution has no higher goal.

Today's Darwinist will reluctantly consider evidence on just about any variation on the theme of evolution—any variation, that is, that is not a result of a purpose. Gould, for instance, writes, "Orthogenesis denotes the claim that evolution proceeds along defined and restricted pathways because internal factors limit and bias variation into specified channels." The *Structure* in Gould's title, *The Structure of Evolutionary Theory*, is a tip of the hat to the *channels* possibility. Today many in evo-devo[191] would agree. Gould, so as not to be excommunicated from Darwinism, however, is careful to add that "all leading orthogeneticists insisted vociferously that their arguments for internal directionality included no teleological or theistic component."[192]

Still, Darwinians have generally condemned anything that smacks of purpose as heresy. Orthogenesis is one example. Lamarck's theory of use and disuse—even though Darwin allowed the theory—was frowned upon because the giraffe's adding longer vertebrae to his neck by effort rather than chance to reach those higher, more succulent leaves and impress giraffe girlfriends came too close to teleology. The idea that consciousness, and its inherent purposes, could be influencing the direction of evolution is anathema among orthodox Darwinians, who demand that consciousness be part of nature, but no part of selection. You may soon find Darwinians on the steps of the New York City Library burning Thomas Nagal's *Mind and Cosmos: Why the Materialist Neo-Darwinian Conception of Nature Is Almost Certainly False.* Nagel is guilty of the ultimate heresy in Darwinism: teleology. He argues that Darwinism, while it may be part of the mechanism for evolution, is insufficient to explain the wonders of genetic material and that something more is needed:

> The teleological hypothesis is that these things may be determined not merely by value-free chemistry and physics but also by something else, namely a cosmic predisposition to the formation of life, consciousness, and the value that is inseparable from them.[193]

However, even though Nagel professes to be a naturalist and a non-theist, even though he is philosophizing about a cosmos without a deity, his secularism is not enough. Darwinism is not just dead set against God. Darwinism is hell-bent on reducing teleology to superstition, on doing away with purpose or goal, natural or supernatural, playing any role in nature. To be admitted to the cult of Darwin is to embrace a faith in a purposeless universe—which brings us to the question of faith.

Faith

Faith has evolved. Consider a passage from a couple of centuries ago on the futility of making faith *respectable* by bolstering it with objective investigation and probability:

> Suppose there is a man who desires to have faith. Let the comedy begin. He desires to obtain faith with the help of objective investigation and what the approximation process of evidential inquiry yields. What happens? With the help of the increment of evidence the absurd[194] is transformed to something else; it becomes probable, it becomes more probable still, it becomes perhaps highly and overwhelmingly probable. Now that there is respectable evidence for the content of his faith, he is ready to believe it, and he prides himself that his faith is not like that of the shoemaker, the tailor, and the simple folk, but comes after a long investigation.[195]

The passage was written by protestant theologian Søren Kierkegaard (1813-1855), who was concerned about an obsession with the evidential, objective Biblical history diluting the experience of transcendence. When it came to a personal relationship with God, which was better: looking inward or at history? In Kierkegaard's words, "No one who has not been corrupted by science can have any doubt in the matter."[196]

The skeptic may suppose that we all have been corrupted by science, or rather by biomythology masquerading as real science. Kierkegaard's assessment of his theological times could easily be applied to Darwin's compulsivity to bolster with evidence his own leap of faith. Making a hypothesis on faith and then searching for data to strengthen the faith has become the *modus operandi* of our time, whether we are speaking of religion or science.

Thomas Kuhn describes the workings of faith in science:

> The man who embraces a new paradigm[197] at an early stage must often do so in defiance of the evidence provided by problem-solving. He must, that is, have faith that the new paradigm will succeed with the many problems that confront it, knowing only that the older paradigm has failed with a few. A decision of that kind can only be made on faith.

Faith at Oxford

So can the skeptic imagine that Darwinians have faith? *Faith* is derived from the Latin *fides*, which shares a base with *fidus* meaning "trustworthy" and *fidere* meaning "to trust." We have faith in that which we trust. Do Darwinians trust Darwin's genius?

According to the *Shorter Oxford English Dictionary*, faith is "confidence or belief." Do Darwinians believe and have confidence in Darwin?

Faith is "belief without … proof." To prove something is to demonstrate it beyond reasonable doubt, but considering how often *reasonable doubt* has been used to execute the innocent, the skeptic prefers—until execution is no longer reasonable—to define faith as "belief without absolute proof." Do Darwinians have such proof? Do they have proof that no future discovery may conflict with their current convictions?

The *Oxford*'s next definition includes "a system of firmly held beliefs." Does Darwinism fit this definition? Is Richard Dawkins, for instance, firm in his beliefs?

Finally, we have "the spiritual apprehension of divine truth or intangible realities." Since few Darwinians have touched an amphibian transforming into a reptile, Darwinism meets the intangible criterion, but what about the *spiritual apprehension*?

Spirituality

Some find their spirituality in casting stones. Some find their spirituality in those "sermons of stone" known as cathedrals. Some find their spiritualty in brimstone. Some search for a spirituality without stones: love of God, neighbor, self, and enemy (the God and self parts often coming easier than the rest of the package).

Spirituality is expanding. There is a new spiritual movement afoot, another rocky road. Michael Shermer provides the description:

> If spirituality means a sense of awe and wonder and humility in the face of the creation, what could be more awesome and wondrous and humbling than the deep space discovered by astronomers and the cosmologists and the deep time discovered by biologists and geologists?[198]

One of the fruits of the spirit is joy, and the joy some followers of science feel when exploring creation would certainly qualify as spirituality. For Shermer, the deep time explored by science evokes the infinite. Imagine the tingle of awe Charles Darwin must have experienced when contemplating the intangible deepness of an omnipotent deep time. Thanks to the incompleteness of the geological record, Darwin's master builder, deep time, is largely untouchable, capable of eliciting "spiritual apprehension of intangible realities."

Then, there is one of the twentieth century's most poetically gifted apologists for science, Carl Sagan:

> Science is not only compatible with spirituality; it is a profound source of spirituality. When we recognize our place in an immensity of light-years and the passage of ages, when we grasp the intricacy, beauty, and subtlety of life, then that soaring feeling, that sense of elation and humility combined, is surely spiritual.[199]

Is creationism science? Is science spirituality? Again, skepticism about words sets in. Can the search for truth really be bounded and divided by such words as science and spirituality? Is there an actual joint between the bones of spirituality and science? The words of Sagan's paragraph capture spirituality; the sense of wonder, joy, transcendence, and profundity Sagan elicits, however, is to be found in the skill of his rhetoric, the lyricism of his prose, and not just the rationality of his science.

Elsewhere[200] Sagan tells us that "every time we exercise self-criticism, every time we test our ideas against the outside world, we are doing science." Sagan's advice to check our ideas about the physical world against the physical world makes sense for the science of electrical outlets, but it falls apart completely when considering the spark of spirituality. The skeptic could imagine that spirituality may be found in the screams of children's laughter on the playground or our love for our neighbors even when they are drunk, destitute, and lying in their own urine on the street, but we need not check those sounds, or that love, against the position of the swing, slide, or curb. The skeptic suspects that we are about as likely to burn a hole in spirituality with the focus of Sagan's objective magnifying glass as we are to answer how many selfish genes can dance on the head of a pin. Yet not only Sagan, but also Richard Dawkins feels the same spiritual call from creation. Dawkins quotes Carl Sagan's *Pale Blue Dot:*

> A religion, old or new, that stressed the magnificence of the Universe as revealed by modern science might be able to draw forth reserves of reverence and awe hardly tapped by the conventional faiths.[201]

Then Dawkins allows:

> All Sagan's books touch the nerve-endings of transcendent wonder that religion monopolized in past centuries. My own books have the same aspiration. Consequently I hear myself often described as a deeply religious man ... But is religion the right word?[202]

A good question. Is *religion* the right word for prose that evokes the wonder that accompanies speculations about creation?

In *Only a Theory: Evolution and the Battle for America's Soul,* author Kenneth Miller best sums up the subject of Darwinism and faith:

> Do we have the strength and the wisdom to allow science to discard the ideas that don't work, and to search for genuine truth about the natural world? To be sure, this requires a certain degree of faith, a faith that there is an objective reality to nature, and the faith that such a reality is indeed worth knowing. There is a risk in embracing faith, even faith in reality, even the faith of a scientist. But faith promises rewards as well, and in finding the strength to embrace what evolution tells us about the nature of reality, we will find reward beyond measure. For it is such faith that will ultimately redeem our scientific souls.[203]

Miller crystalizes the essence of his story about science (and truth): what works. The difference between mere faith, mere biomythology, and real science rests on what works to control nature. Whatever the distractions of evidence, the skeptic doubts but that we take our cosmologies on faith, and there can be little doubt that Darwinism depends on faith: some trust it, some accept it without absolute proof, some hold it as a system to be firmly believed until the next discovery or revelation, and some find it downright spiritual in a naturally aesthetic sort of a way. Even though Darwinism requires faith, does it qualify as a religion?

Praise Darwin

What is religion? For most in the West (Richard Dawkins being no exception), religion is defined as "belief in God." This definition, of course, eliminates two of the five major world religions. The Buddha said little about any god, and Hindus have more gods than Eli Lilly has antidepressants. That many Darwinists are atheists provides no evidence against their religious yearnings. Many Buddhists devote their lives to their religion and yet claim no ties to theism. Possibly with this discrepancy in mind, the 1995 *HarperCollins Dictionary of Religion* looks past the "religion is belief in God" definition to offer as an "adequate" definition: "One may clarify the term religion by defining it as a system of beliefs and practices that are relative to superhuman beings," which the definition goes on to describe:

> Superhuman beings are beings who can do things ordinary mortals cannot do. They are known for the miraculous deed and powers that set them apart from humans ... Furthermore, the definition requires that such superhuman beings be specifically related to beliefs and practices, myths and rituals.[204]

At first glance, this definition seems to exclude Darwinism as a religion—but only at first glance. The scientific method is certainly a ritual, and gradualism may soon be acknowledged as a myth. But what makes biomythologists "superhuman beings"?

Modern science is generally dated from the "The Age of Reason." Again and again, we are told that science is based on reason and not emotion. On the Internet, Richard Dawkins boasts a

"Foundation for Reason and Science," and Adam Gopnik's previously quoted words agree that "this rootedness in reasoning explains why of all explanations of life, evolutionary theory is not remotely like a religion." It seems that biomythologists work off reason, not emotion. Ironically, this is exactly what defines them as superhuman beings.

The Science of Superhuman Beings

Consider the work of Antonio Damasio. *The New York Times* rates Damasio as "one of the world's leading neurologists." This I know, for the cover of his *Descartes's Error: Emotion, Reason, and the Human Brain* told me so. Setting out, perhaps unwittingly, to prove Hume's dictum—"Reason is and ought only to be the slave of the passions"—Damasio, in his beautifully crafted prose, gives examples of how those who have suffered damage to the emotional centers of the brain cannot make decisions, reasonable or otherwise. In his preface, Damasio writes, "When emotion is entirely left out of the reasoning picture, as happens in certain neurological conditions, reason turns out to be even more flawed than when emotion plays bad tricks on our decisions."[205]

In "A Passion for Reasoning," his concluding chapter, Damasio repeats: "At the beginning of this book I suggested that feelings are a powerful influence on reason, that the brain systems required by the former are enmeshed in those needed by the latter, and that such specific systems are interwoven with those which regulate the body."

To be human, as Damasio recounts it, is to be unable to make reasonable decisions without emotion. If Damasio is correct, when a Darwinian claims the ability to divorce reason from emotion ("Just the facts, ma'am; just the facts"), the Darwinian is declaring himself to be a superhuman being and qualifying Darwinism, with its beliefs and practices revolving around superhuman beings, as a religion.

The skeptic suspects that if biomythology is correct about what makes us human, either Darwinism is inextricably tied to emotion or it is a religion—take your pick.[206]

If practice makes perfect, when does a practice become a religion? In the words of Kierkegaard, "Take passion away and faith disappears."[207] In the words of Protestant theologian Paul Tillich (1886-1965), faith is "ultimate concern":

> If it claims ultimacy it demands the total surrender of him who accepts this claim, and it promises total fulfillment even if all other claims have to be subjected to it or rejected in its name.[208]

To the mind of the skeptic, we worship the mundane in many guises: money, sex, fame, sports. Ask anyone whose spouse has been lost to the cult of Monday Night Football. The passion and ultimate concern of philosopher Daniel Dennett, however, transcends the mundane:

> Evolution ... is the central, enabling process not only of life but also of knowledge and learning and understanding. If you attempt to make sense of the

world of ideas and meanings, free will and morality, art and science and even philosophy itself without a sound and quite detailed knowledge of evolution, you have one hand tied behind your back.[209]

What concern could be more ultimate than "other claims have to be subjected to it or rejected in its name"? From the perspectives of Kierkegaard and Tillich, let's visit or revisit a bit of the passion and ultimate concern, if not the downright religious enthusiasm, of Darwin's disciples for their prophet's genius:

Richard Dawkins:

Fred Hoyle was a brilliant physicist and cosmologist, but ... he needed to have his consciousness raised by some good exposure to the world of natural selection. At an intellectual level, I suppose, he understood natural selection. But perhaps you need to be steeped in natural selection, immersed in it, swim about it, before you can truly appreciate its power.[210]

Steven Pinker:

Natural selection has a special place in science because it alone explains what makes life special.[211]

Edward O. Wilson:

The great questions—"Who are we?" "Where did we come from?" "Why are we here?—can be answered only, if ever, in the light of scientifically based evolutionary thought.[212]

Kenneth Miller:

> We are among those organisms [evolution's winners],
> but we possess something special, something that no
> other organism to our knowledge has ever had—the
> ability to see and to understand how we came to be.
> This is a precious gift, and we must never lose it.
>
> The elegant universe is a universe of life. And the
> name of the grand design of life is evolution.[213]

Jerry Coyne:

> But there is something even more wondrous. We
> are the one creature to whom natural selection has
> bequeathed a brain complex enough to comprehend
> the laws that govern the universe. And we should be
> proud that we are the only species that has figured
> out how we came to be.[214]

Michael Shermer:

> Darwin matters because evolution matters. Evolution
> matters because science matters. Science matters
> because it is the preeminent story of our age, an
> epic saga about who we are, where we came from,
> and where we are going.[215]

Daniel Dennett:

> If I were to give an award for the single best idea
> anyone has ever had, I'd give it to Darwin, ahead of
> Newton and Einstein and everyone else. In a single
> stroke, the idea of evolution by natural selection
> unifies the realm of life, meaning, and purpose with
> the realm of space and time, cause and effect,
> mechanism and physical law.[216]

So what does the cosmology of Darwinism teach the skeptic about how we rose above the rank and file of primates? Apparently what defines us as not mere animals, but human animals, is to experience the "raised consciousness" afforded by "the single best idea anyone ever had," to be like "no other organism," "the only species," "the one creature" to receive the "precious gift" of knowledge that our lives were "bequeathed" by natural selection, that the "grand design of life is evolution," that evolution is "the central, enabling process not only of life but also of knowledge and learning and understanding."

If a belief inspires proselytizing, evokes a sense of transcendence and spirituality, demands ritual and reverence for consensus, describes our origins and starless afterlives, and promises certain truth despite uncertain knowledge, then is it religion? If someone quacks like a zealot, waddles like a zealot, and swims like a zealot, should the skeptic still duck?

Apostasy

In the early Catholic Church, the list of mortal sins included adultery, murder, and apostasy—the act of denying one's faith. Thus when a Christian was given a choice between being fed to the lions of recanting his faith, he was forced to make a choice—this life, or the next. The dilemma encouraged many a deathbed baptism. The Emperor Constantine, for instance, being responsible for his share of deaths as he united an empire, was no fool. He waited to the very end for the baptism that would wash away his sins.

Today, if materialists admit that Darwinism is a faith, they stand to lose the status of their superhuman, emotionless rationality. Even so, the skeptic suspects that apostasy is apostasy. Whether we worship God or Darwin, we cannot *know* what future reasoning and discoveries will reveal. Therefore, we must take the truth of our stories—whether revealed, discovered, or evolved—on faith: certainty that mystery's weight loss program will never reveal the ribs of a contradiction in our propositions, reasoning, or beliefs.

Chapter 8

The Rhetoric of Objectivity

Having enhanced our doubts to include the father of biomythology, Charles Darwin, and his theory, our primer for skeptics will now study Darwin's method of selling: the science of persuasion, also herein called *the rhetoric of objectivity*. This approach, under both names, has been included in our straw-man definitions of biomythology so that these chapters could handily be included in this polemic on the virtues of skepticism.

In the *Shorter Oxford English Dictionary*, one definition of *rhetoric* is "the art of using language so as to persuade or influence others." Sometimes the word implies eloquence stealing our breath and reason away; sometimes the word implies substituting skill and cleverness for sincerity. Over two thousand years ago, Aristotle wrote a book on the subject. He divided persuasion into three modes: character, emotion, and proof or apparent proof.[217] Of character, he wrote that "his character may almost be called the most effective means of persuasion he possesses." Of emotion, Aristotle explained that "persuasion may come through the hearers, when the speech stirs their emotions." Of proof, Aristotle concluded that "persuasion is effected through speech itself when we have proved a truth or an apparent truth by means of the persuasive arguments suitable to the case in question."

For the purposes of understanding, it is helpful to divide and conquer. When I describe vision, for instance, I speak of the ciliary muscles inside the eye that refocus the lens to keep things clear; I describe the extraocular muscles that aim the eyes to fuse two disparate views of the world into a single three-dimensional image. In truth, seeing is a unified act, more than calisthenics, more than the sum of its parts.

Are Aristotle's three modes really separate? Is he cutting persuasion at the joints or, blinded by his love of wisdom, butchering reality? First, character is a quality used to earn trust. Honesty, dependability, loyalty, competence, erudition—all enter into it. Trust is nevertheless an emotion. Secondly, eloquence or artifice stirring emotion obviously relies on emotion. Thirdly, presenting a brilliant proof can generate the emotions of awe, admiration, and respect in those conditioned to revere reason (Philosophy is the love of wisdom; love is blind; therefore, philosophy is ... Equivocation on the word *love*, maybe, but medieval logic was the invention of those reconciling a love of reason and God). Thus, none of persuasion's three modes muzzles the bite of passion. Passion, after all, is what turns a story into a belief. Darwinians and creationists are passionate about their stories and passionate against stories that oppose their own.

One trick in the rhetoric of objectivity is to create a passion for objectivity, to remind us constantly of our debt to rationality. Those indoctrinated in this passion by those they respect and admire easily cling to apparent truth long after reason has moved on to its next game. Objective arguments appeal to the emotion of those who cherish the story of reason.

Even before Aristotle, itinerant teachers known as Sophists—from the Greek noun *sophia*, meaning "wisdom" or "learning"—taught the art of persuasion. In a democracy such as Athens, the art of persuasion was key to political success. In the America of today, nothing has changed. What has changed is the injection of modern science into the art of persuasion.

In the case of the rhetoric of objectivity, aligning one's name with the term *scientific* bolsters character and trust. Add this to the emotion of gratitude for real science bringing us and our loved ones decades nearer to immortality, as we no longer die from toothaches or smallpox, and we easily forgive theory's occasional lapses, using ad hoc arguments of "because, because, because,"

As I defined the terms in earlier chapters, real science is the search for truth; biomythology, the sale of truth. Biomythology, per this definition, is thus rhetoric. Using natural selection rather than biochemistry, for instance, to evolve a new bactericidal agent—if it could be done—would be an example of real science. Offering the particulars of bacterial mutation to sell the truth of microbes evolving into marching bands is biomythology, an example of the use of science to persuade.

As discussed, real science has changed the physical world (and the mental world if we allow that comfort, health, full bellies, freedom from the elements, and a longer lifespan have improved our frame of minds). The word *scientific* has thus become an emotionally charged sales tool. The thought of betraying science or having our smartphones confiscated because we fail to believe in natural selection is an excellent way to stir the emotions. Science, we are told, is based on objectivity, not subjectivity, as if the Law of Gravity came from the objectivity of the apple, not the subjectivity inside Newton's head.

Thus, the rhetoric of objectivity includes the art of concealing subjectivity. Intelligent designs, evolutionary equivocation, substituting the goal of materialism for the tool of naturalism, turning the philosophy of instrumentalism into a crime and Darwin into an object of veneration—all are examples of the devices of the rhetoric of objectivity. The "ghost in the prose" is another.

The Ghost in the Prose

Take the word *I*. Just as British philosopher Gilbert Ryle (1900-1976)—equating eternity with poor taste—lampooned the soul as "the dogma of the ghost in the machine"[218] known to the rest of us as "I," the rhetoric of objectivity lampoons the *I* as the ghost in the prose better known to readers of science as the author. The ghost supposedly enjoys, to borrow the words of philosopher Thomas Nagel's title, *The View From Nowhere*. Substituting *the author* for *I* can transform any purveyor of biomythological goods into a paragon of reason and empirical wisdom. Neuroscientist Christof Koch notes how this new rhetoric has captured the minds, hearts, and pages of the scientific community:

> I also write in the face of a powerful professional edict against bringing in subjective, personal factors. This taboo is why scientific papers are penned in the desiccated third person: 'It has been shown that ...' Anything to avoid the implication that research is done by flesh-and-blood creatures with less than pristine motivations and desires.[219]

The skeptic doubts but that just as magicians use the flourish of cape to distract the eye, so the rhetoric of objectivity uses the flourish of thousands of hours of meticulous data collection and exacting calculation to distract journal review boards from interpretations tailored to fit passion and from samples selected as cleverly as juries. Like magicians, biomythologists fool their audiences. Unlike magicians, biomythologists too often fool themselves.

Plain Prose

Plain prose is another device that every skeptic needs to recognize. The device is used in the rhetoric of objectivity to hide subjectivity. Renaissance humanism began the search to unearth the classics and their breed of older wisdom. The movement inspired others to demand to see the Bible itself, not just catechisms and Church doctrines. Studying rhetoric, the language and persuasive techniques of classical Rome and Greece, was also stressed so the recovered wisdom could be sold and society bettered. The Renaissance, the skeptic might allow, was built on wisdom and eloquence. The Enlightenment jettisoned both for facts and progress—not always a good idea. Facts are like speed, but wisdom is like velocity—it includes direction. Progress can too easily be aimed in the direction of terror.

Just as in business, *flat* is the new *up*, so when it comes to objectivity, plain prose is the new rhetoric. Back when natural philosophy was first turning up the lights to abolish the shadows of superstition, Bishop Thomas Sprat (1635-1713) in his *History of the Royal Society* began our salvation from rhetorical extravagance:

> Thus they have directed, judged, conjectured upon and improved experiments. But lastly, in these and other businesses that have come under their care, there is one thing more about which the Society has been most solicitous, and that is the manner of their discourse; which, unless they have been very watchful to keep in due temper, the whole spirit and vigor of their design had been soon eaten out by the luxury and redundance of speech. The ill effects of the superfluity of talking have already overwhelmed most other arts and professions, insomuch that when I consider ... the causes of their corruption, I can hardly forbear recanting what I said before, and concluding that eloquence ought to be banished out of all civil societies as a thing fatal to peace and good manners.[220]

Thus Sprat eloquently banished eloquence from England's budding academy of science. He traded one virtuoso performance for another, one passion for another: our awe at the aesthetics of prose for our awe at the aesthetics of winning reasoning and observations. Passion for the virtuosity of performance continues to make the sale. Just as the brilliance of the sermon or violin concerto or field goal may be the closest some will ever come to transcendence, the magic of an undeniable thought experiment or argument with surprisingly perfect examples inspires us to believe, at least until time and creativity provide the skeptic with a new perspective. Then we see how the trick was performed.

Eloquence, poetry, aesthetics, philosophy, novel examples and novel thought—all startle us from blind habit long enough to open our consciousness. The feeling evoked is as sweet as a first peek at a true love or dream car. The expansion of consciousness is the truth, not necessarily the proof, provided by the prose or philosophy. Can truth be found in the mere arrangement of facts? Hoping this is true, today's writers, in all areas, have adopted the Royal Academy's plain prose as if to proclaim, "We got evidence; we got objectivity; we got freedom from superstition!" Only a skeptic might suppose that science succeeding with mind as it succeeded with matter may itself be a superstition. Novelists, even some poets, cast aside their lyricism adapting plain prose and plain poetry to sell what often amounts to plain nonsense.

Despite such trends, Darwin hardly let the restrictions of Sprat get in his way. Consider the poetry of his prose:

> Thus from the war of nature, from famine and death, the most exalted object which we are capable of conceiving, namely, the production of the higher animals, directly follows. There is a grandeur in this view of life, with its several powers, having been originally breathed into a few forms or into one; and that, whilst this planet has gone cycling on according to the fixed law of gravity, from so simple a beginning endless forms most beautiful and most wonderful have been, and are being evolved.[221]

Darwin's words seduce us as might the beauty of autumn leaves eddying upward into trees of fire: the beauty and the topsy-turvy viewpoint of predicting the past. *What Darwin Got Wrong* details the confusion:

> History (natural history included) is about what actually happened; it's not about what *had* to happen, or even about what would happen if Mother Nature were to try again. What had to happen is the domain of theory, not of history; and there isn't any theory of evolution.[222]

Darwin may have reversed the arrow of time, but he knew how to use character, prose, emotion (of a new discovery), facts, and reason to sell a new cosmology.

Cook's Law: Rhetoric Equals Mind over Matter

So when is science real science, and when it is rhetoric? In optometry, rhetoric is largely unnecessary when claiming an exact lens can burn a new navel in a fly at exactly one distance. Such optical vivisection qualifies as real science because it is repeatable— for the sport of wanton boys and wanton researchers alike. Rhetoric takes over only when we try to predict how the fly feels about the sting of the smoke in his eyes, or how a particular patient and optics will bond. Scientific certainty and repeatability vanish the moment consciousness enters the equation.

To aid the skeptic, I have transformed these observations into a simple equation. Call it Cook's Law, if you will:

$$Rhetoric = \frac{Guesswork \times Mind}{Matter}$$

The equation tells us the more mind, the more rhetoric; the greater the control of matter, the less rhetoric; the greater the number of guesswork words—such as *maybe, might, could, possibly, most probably, evidently, reasonably, almost certainly, suggest*s, or the number of extrapolations, estimations, inferences and unsolved questions—the more rhetoric. Flipping a light switch requires no rhetoric; flipping a mind requires little else. Perhaps this is why the American Academy of Optometry is dedicated to the "art and science" of vision. The science is in how lenses bend light. The art is in how light bends the human eye and heart.

So what does the equation tell the skeptic about biomythology? What separates us from animals—or vegetables, depending on our definition of *animal?* It's not our eyesight that makes us human;

some creatures see better, some worse. What makes us human is those stories, constructed of words and pictures, we call our minds. That humans are animals is a wonderful metaphor when applied to biology; it becomes biomythology when applied to living our lives. The misapplication rivals applying Newtonian physics to atomic particles. Universals extend only to the far reaches of their own universes. Thus, per our equation, real science elevates the quantity of rhetoric when it crosses the line to leave the common denominator of matter and rise into the numerator as mind.

Is there any support for the equation? In the April 25, 1953, issue of *Nature*, an article appeared: "A structure for deoxyribose nucleic acids." Written by James Watson and Francis Crick, the article was probably the most famous addition to biological science to appear in the twentieth century. The article eventually led to a dramatic increase in our control of nature, even though it was but little longer than a page. Darwin's *Origin* exceeds 500 pages, and natural selection gives us little control over anything but ideology. Per our equation, which piece of writing do you suppose contains the most rhetoric? Granted, both pieces are considered part of the scientific literature, but which one belongs to real science and which to biomythology?

Darwin's Science of Persuasion

Darwin invented scientific rhetoric no more than he invented evolution. He merely packaged both for general consumption. It was Darwin whose rhetoric first sold evolution to scientists and the public. Rather than asking if anyone wanted to buy evolution, *Origin* asked with professional sales acumen, "Which color would you prefer, 'natural selection' or 'use and disuse'"? Ever since, buyers have wondered about the choice of color, but their signatures were already dry on the bill of sale. Darwin deserves to be remembered for the master rhetorician and storyteller he was. He created a tale that continues to sell, and amuse, almost two centuries later.

The Analogy

The false analogy, as any skeptic might suppose, has been a rhetorical device ever since the pre-Socratics began their quest to dignify myth with the pretensions of philosophy. Analogies became the new myths. Extended comparisons of two things became sought after and admired. For example, "A ball is round like an orange: it rolls, it bounces, etc." Just don't expect orange juice.

To the ear of the skeptic, "false analogy" is redundant. Analogies, while true from one perspective or many, are most always false from another; otherwise, they would be identities, not analogies. Without myths, analogies are typically about as good as we can do. In the physical world, the quarks won't stand still long enough for identities. Unless you are able to transcend the physical, you are but a mere analogy of your former self, a mere analogy of your future self. Persuasion is the art of selling analogy as identity. If the skeptic will excuse my analogies, analogies are like the façades of buildings on a movie set propped up by the two-by-fours of conflicting perspectives in the back. The façade's illusion depends on the freezing of perspective by the camera—or fMRI. The art of persuasion is in creating a façade that is so aesthetic, so novel, so riveting, that no one thinks to look behind it. Soon the façade congeals into fact.

Conflicting perspective is the death of analogy. From the perspective of a skeptic, analogies are like appearances: not until viewed from every perspective, inside and out, will truth emerge—a truth that is greater than the sum of the perspectives. The instruments of real science provide new perspectives, routinely crumbling old analogies. Biomythology thrives best on the paucity of perspective.

Questions spur the search for new perspectives; answers deny them. The difference between philosophy and biomythology is that philosophers never tire of looking for new perspectives. This is why philosophy provides more questions than answers, while biomythology confuses analogy with identity to substitute questionable answers for truth. To repeat, every good skeptic doubts but that analogies (though wonderfully persuasive tools) are mere rhetoric. That the pieces between those things being compared enjoy a felicitous fit hardly qualifies them as truth. A myth is a lie that tells the truth. An analogy is a myth waiting to be exposed by a new perspective.

That said, Darwin begins *Origin* with a wonderfully persuasive analogy, which like all analogies crumbles when viewed from other perspectives. In his first chapter, "Variation under Domestication," he develops the concept of artificial selection, the way animals are bred by humans. There are many kinds of dogs, for instance. Darwin then personifies Mother Nature, turning her into a person who has bred the plants and animals of the world.

To the skeptic, the analogy has its problems. The anthropomorphic Mother Nature, it turns out, is more powerful than other humans are. To perform her miracles, she has the ever faithful, omnipotent Deep Time at her side. While mere humans, no matter how long they breed dogs, never come up with cats, and cats, however much they are bred, never show the variations of dogs,[223] deep time—by tiny, gradual steps—can breed reptiles out of amphibians and mammals out of reptiles one tiny biomechanical step at a time.

Another problem is that according to today's perspective humans are just animals, another part of nature. If humans, therefore, select for a breed of dog, nature selects for a breed of dog. There is nothing artificial about artificial selection. *But humans have minds,* you say. *You said it yourself.* Yes, it occasionally happens, but those minds, we are told, were selected. Unless this is an example of unnatural selection, or unless those minds are somehow supernatural—as many have supposed—then whatever is selected by those minds is just another pale example of natural selection.

Either humans are fully natural or they transcend nature (meaning they are not fully natural). The skeptic doubts we can have it both ways by a trick of language. If beehives and beaver dams—produced by animals—are products of nature, and if humans are merely animals, then whatever we create is natural: babies, smartphones, atomic bombs, global warming. Take your pick. This, of course, is a false alternative. Perhaps just as things are true in science except for when they are not, nature is nature—except for when it isn't? Or, perhaps our term *nature*, which excludes the products of the natural mind, has been artificially selected by popular vote; perhaps, as we will discuss,[224] in dividing what exists, we have cut along the wrong dotted line?

The skeptic may be quick to note that from other perspectives, my analogy is obviously false, but that is okay. The analogy is no falser than Darwin's.[225] The point is that analogies are seldom more than rhetorical devices, and Darwin has used his prodigious knowledge of nature to provide a vivid analogy for selling his theory. Viewed from the correct angle, a banana is shaped like a rocket, but an analogy, however brilliant, won't get the skeptic to the moon.

The Elephant We Never Forget

In rhetoric, there is nothing quite like a good mental image to make a sale. Take Darwin's elephant example:

> The elephant is reckoned to be the slowest breeder of all known animals, and I have taken some pains to estimate its probable minimum rate of natural increase: it will be under the mark to assume that it breeds when thirty years old, and goes on breeding till ninety years old, bringing forth three pair of young in this interval; if this be so, at the end of the fifth century there would be alive fifteen million elephants, descended from the first pair.[226]

This example packs even the skeptic's mind with unforgettable never-forgetting images. Providing that the first pair of elephants were not more interested in trumpet duets than breeding, the math works out, I'm told, turning the theory into science, a science that completely failed to predict that the advent of ivory hunters, not pachydermal capitalistic competition for a limited number of peanuts, would determine the fate of the species. When it comes to elephants, the race was hardly given to the swift. The fitness was in the fit with the environment, not the fitness of the bulls. It was the aim of the hunters that counted, not the tap-dance competition between elephants. The biggest and strongest bull elephants, sporting the most ivory, were probably shot first. Again, if man is merely a part of nature—as we are told—then whatever man selects is natural selection, period. Still, just as Darwin intended, the image of procreating elephants filling the world sticks in our minds, even when the number of elephants in the world is decreasing, not increasing as the rhetorical math of Darwin and the right Reverend Malthus predicted.

The Caveat

A caveat is a warning or caution. Every good skeptic should be aware of the device as it is used in the rhetoric of objectivity. By law, for instance, television pharmaceutical commercials always end with a voice (its heightened speed and pitch seemingly placing it in parenthesis) offering a caveat about the possibility of the advertised drug causing leprosy, dysentery, or dyslexia. Similarly, when we call our insurance companies, the same parenthetic voice offers a caveat that anything the representative promises may be a blatant lie, and if the elective procedure doesn't kill us, the bill might.

Caveats are nothing new. As early as 1620, Sir Francis Bacon advised aspiring natural philosophers to "sprinkle warnings, reservations and cautions in all directions."[227] It was Darwin, the skeptic might suppose, who elevated the rank of the caveat in the rhetoric of objectivity. In 1859, he concluded the introduction to *Origin* with, "I am convinced that Natural Selection has been the main but not exclusive means of modification." This allowed Darwin to display "Natural Selection" prominently in his title, devote thousands of words to his theory, but declare his open-mindedness and include every other possibility with the sweep of a single sentence. This alerts the skeptic because the device continues to

---

David Cook

allow Darwin's disciples to claim all descent with modification for the turf of their prophet even though he said relatively nothing about other modifications. Natural selection can easily be reduced to, "Nature, not God, did it."

Gould offers references on how Darwin protested when his caveat was ignored:

> As my conclusions have lately been much misrepresented, and it has been stated that I attribute the modification of species exclusively to natural selection, I may be permitted to remark that ... I placed in a most conspicuous position—namely at the close of the Introduction—the following words: 'I am convinced that natural selection has been the main, but not the exclusive means of modification.' This has been of no avail. Great is the power of steady misinterpretation.[228]

In the same footnote on the same page, Gould quotes from a 1900 essay by Darwin's friend, G. J. Romanes, "In the whole range of Darwin's writings there cannot be found a passage so strongly worded as this: it presents the only note of bitterness in all the thousands of pages which he has published."

Darwin's righteous indignation aside, the caveat remains a favorite device in the rhetoric of objectivity. Michael Shermer warns of "provisional truths," truths that future discoveries may prove wrong. Sam Harris warns us, "There is no question that scientists have occasionally demonstrated sexist and racist biases."[229] We are reassured to know that Harris is objectively considering all perspectives. Jerry Coyne warns, "As we will see, it is possible that despite thousands of observations that support Darwinism, new data might show it to be wrong."[230] Perhaps this should have been the subtitle of *Evolution is True*?

The rhetoric of objectivity thrives on caveats, even those rhetorically softened by "thousands of observations." We will return to Coyne's warning when we examine the "uncertain knowledge" that goes by the name of scientific truth, but at this point we are discussing the device of the caveat in the rhetoric of objectivity. As we saw in our introduction, *scientific* position statements sell what they are selling in the abstracts and bury the caveats deep in the body of the paper, where they will not interfere with the sale but can, nevertheless, be dug up should proof of objectivity be required. In real science, the caveats need not be hidden: the write-up for Watson and Crick's monumental discovery begins with the words, "We wish to suggest a structure for the salt of deoxyribose nucleic acid (DNA)." The author's use of the word *suggest* is a caveat, a warning, or admission that they are just making a suggestion and could be wrong. As luck would have it, this caveat—for now, at least—turned out to be one of the least needed of the century. Still, it never hurts to play it safe. Caveats—look for them wherever ideas are being sold.

The Silence of Jets

Righteous indignation aside, Darwin selected for a more successful variation of the caveat that every skeptic should learn to recognize. Nothing blinds us to distracting peripheral observations more quickly than simply acknowledging them. Suppose, for example, I'm lecturing on the difference between hacksaws and sawdust, and the roar of a jet flying overhead distracts the audience. The simple statement "Boy, that jet is loud!" can easily end the exploration of the roar and return the audience's attention to my old saws— whether hack or dust. Should we bother to listen, the roar of the jet would remain as loud as ever, but our minds are now silenced. I have dubbed this rhetorical device "The Silence of Jets." It appears throughout Darwin's "Difficulties of Theory" chapter.

Darwin asks, "Can we believe that natural selection could produce ... organs of such wonderful structure, as the eye, of which we hardly as yet fully understand the inimitable perfection?"[231] He allows 900 words (five paragraphs) to removing the miracle from the creation of the eagle's eye.[232] A third of these words he devotes to silencing the scream of jet engines, to deafen us to noise seemingly at odds with his proposed mechanism for evolution. He notes the absurdity of his theory applied to the complexity of the eye, the lack of lineal ancestors to support the theory, the imperfect fossil record concealing the transitions, and the natural temptation to turn to the Creator to explain the miracle. Only when we are oblivious to the scream of the jets, and Darwin has our undivided attention, does he use his prodigious command of the minutiae of natural history to embellish his explanation of how natural selection could have fashioned the eye. He employs the device repeatedly. "Firstly, why, if species have descended from other species by insensibly fine gradations, do we not everywhere see innumerable transitional forms?"[233] And, "It has been asked by the opponents of such views as I hold, how, for instance, a land carnivorous animal could have been converted into one with aquatic habits; for how could the animal in its transitional state have subsisted?"[234]

These are all excellent questions. Once asked the discrepancies are muffled, heard no more than the roar of those jet engines. Darwin's devotees still use the device. In 2013, after allowing that "natural selection automatically conserves *whatever has worked up to now*,"[235] and "all the dumb mistakes [of evolution] tend to be invisible, so all we see is a stupendous string of triumphs,"[236] Dennett has to take care of the roar of the dinosaurs whose sudden demise by comet seems to contradict his statements. He therefore dismisses catastrophic selection with the admission, "It has always been obvious that the most perfect dinosaur will succumb if a comet strikes its homeland with a force hundreds of times greater than all the hydrogen bombs ever made."[237] So what happened to the marvelous process that "automatically conserves"? Does it conserve except for when it doesn't? What happened to the string of triumphs of natural selection? Was the comet unnatural? Did it not select?

The jet engines continue to roar. The Darwinists, who write high-school science texts, are apparently no longer aware, but the proponents of structuralism and evolutionary development and intelligent design continue to listen to the racket, not the rhetoric.

Name Dropping

In the rhetoric of objectivity, Darwin proves that, as always, it's not just *what* you know; it's *whom* you know. In the four-paragraph preface of the original edition of *The Voyage of the Beagle*, Darwin dropped the names of Captains Fitz Roy and Beauford, the "surgeon of the *Beagle*, Mr. Bynoe," the Reverends Professor Henslow and J. M. Berkeley, and ten others connected to Darwin or his work. Even more impressively, he associated himself with "the Lords of the Admiralty," "the officers of the *Beagle*," "the Lord Commissioners of Her Majesty's Treasury," and "the Right Honorable Chancellor of the Exchequer." Darwin left no doubt that he was a member of the who's who of his day. In *Origin*, he dropped the highly respected Sir Charles Lyell's name twice by the end of the second page of the introduction. Before Darwin was finished, he had carefully, and favorably, referenced as many naturalists as possible who might one day review or promote his book.

While Darwin dropped names, his disciples drop footnotes. Sam Harris's *The Moral Landscape: How Science Can Determine Human Values* and Steven Pinker's *How the Mind Works* each provide about 800 references—Sam a little less, Steve a little more. However, are the references used merely to credit others, or are they provided to establish the author's erudition? Sam, for instance, appears to include everything he had read for his doctoral dissertation, not just the references needed to directly support his text. Pages of footnotes and references are quite impressive and very much a part of the rhetoric of objectivity. Nothing sells quite like a reference. Whether the reference is germane is not the point. Who can be down on an author who is up on the literature? Why else, the skeptic might wonder, would my polemic include over 400 mostly useless footnotes, many of them footnoting the very book we are reading?

Science and Literary Criticism

As we argued in our introduction, *scientific* has become the cliché of clichés. Everyone wants to be *scientific*. Today in literary criticism, there are those who would analyze literature under the lens of Darwinian science, just like former generations analyzed literature under the lens of Freudian and Marxian science, or whatever science was in vogue at the time. What lens will we use to analyze tomorrow's literature? As the skeptic knows, we can't be certain, but digital science is a good bet, which brings us to another approach for making literary criticism *scientific*.

We can, for example, feed the words of a story into a computer and let the computer discover for us the latent patterns. We can have the computer compare the responses of 100 subjects who read a story to that of a 100-subject control group who did not read the story. The control group would, of course be essential because nonreaders and readers alike are often equally opinionated as to whether a given book should be banned or burned.

To make the study more scientific, such opinions could be ignored and the subjects could be food-deprived and tested by gastric secretions, as were Pavlov's dogs. Combining the best of Darwinism, behaviorism, and computer analysis, 100 apes could view the same pages and their gastric secretions be subjected to cross-species analysis and correlated with their genomes. Of course, care would need to be taken so the apes would not eat the books.

We do not have enough room to explore all the other possibilities for making literary criticism *scientific*. The skeptic might suppose, though, the reduction of literary criticism to science would steal away the soul of literary criticism. How authors allow us to abandon routine seeing long enough to envision ourselves in novel ways, how they expand our empathy for those we previously saw as stereotypes, how they make us conscious, how they evoke or bury the transcendence of what it means to be human in a changing world—all would be lost. Literary criticism and science, however, need not be strange bedfellows, but rather than a science of literary criticism, we need a literary criticism of science.

Such literary criticism applies especially to biomythology, that confiscation of scientific explanation to engineer belief. In many of the essays that follow, we will examine other devices used in the rhetoric of objectivity: reason, logic, observation, and evidence, to name a few. If the rhetoric of objectivity is the art of concealing subjectivity, then the job of literary criticism, and skepticism, is to read between the lines to discover that art.

Chapter 9

Musical Definitions

It is as if I were to say: 'You surely know what 'It is 5 o'clock here' means; so you also know what 'It's 5 o'clock on the sun' means.
—Ludwig Wittgenstein[239]

What's in a name? *Monster? Hero?* A serial killer by any other name is still a serial killer. A name can hardly relieve your grief if your own son or daughter becomes part of the series.[240] The skeptic, therefore, has every right to doubt names, every right to doubt words.

For years married bachelors were forbidden in better logic everywhere. Now bachelors are married every day and never kiss a bride. Words won't hold still. Their movement defies logic, but never passion. Much as bodies are the basic units of murder, words are the basic units of speech and writing: you may live or die by them. You may hang on words or be hung for uttering the wrong ones. You may play on words, stand on words, find words or be at a loss for them. Words may be sound or noise; your spoken words, sound and unsound; your written words, sound but not sound—all without contradiction. You may send word or receive word, waste or weigh words, put in a good word, or share a word with the wise. Your word may be taken or given; you may leave word, take my word, or give your word. You can be word deaf, word blind, or just plain dumb. You can put words into or take the words out of a person's mouth. You can eat your words even when they are dirty, foul, or disgusting. You can be as good as your word, whether your words are empty or not. You can put in a good word for or have a word with. You can waste or weigh words that rhyme or reason or pun or punish. Your words may fit the occasion, or you may not be able to get a word in edgewise. Getting in the last word is to be admired; uttering your last word is not. Such is the nature of wordplay, or if you would rather, rhetoric.

Words are not all play. Words can be work, especially when they fail to work. Words can be of doubtful value, especially when doubt fails. As mentioned in the introduction,[241] twentieth-century governments used the words of well-reasoned ideologies, often backed by *scientific* theories, to justify the murders of over 169 million human beings, not counting the millions who died in *scientific* warfare. Even though actions, not words, kill people, and people, not words, tell lies, words too often inspire actions from the skeptically deficient. As long as persons are unschooled to doubt words, words matter. Used in descriptions of the present, the word *progress* means, "an experiment altering habits of language, perception, and action." Used in descriptions of the past, the word *progress* means, "The experiment worked." If the experiment did not work, the word *atrocity* or *terror* may be substituted. Therefore, it behooves the skeptic to recognize the many ways in which the liquidity of language can fuel rhetoric. To help accomplish this, we continue to visit the devices used in the rhetoric of objectivity—this time musical definitions. Musical definitions, in the skeptic's mind, should be those that can be changed as necessary to keep a theory or cash register ringing.

Noon on the Sun

In our chapter "Evolutionary Equivocation," we began our exploration of the language pitfalls that can be exploited to rescue any dying theory from the rigors of too much observation, discovery, or imagination. In this chapter, we will continue that exploration, prefacing our explorations and skepticism with a thought experiment—an eye game, if you will, or a language game, if you would rather.

Picture yourself standing barefoot in a pasture at high noon with the burning slice of freshly-sharpened blades of green, green grass masquerading as cool beneath your toes while the starry, starry noonday sun burns overhead just as high as high can be—as high as the best definitions of *high noon* command.

Got it—cool grass, warm sun? Hold it. Hold it. Good.

Now instead, imagine you have taken an eight-light-second leap to stand on the sun itself—that fiery orb smack dab in the center of Copernicus's tall tale about the heliocentric universe. Again, it is high noon. Where must you now stand to project the noonday sun as high overhead as high can be? What out-of-body contortions and calisthenics must you endure? Were you to stand on your head on the sun's south pole, would the sun now qualify as directly overhead? As the topsy-turvy thermonuclear maelstrom simmered the moral law inside you, would not solar gravity quibble or disclaim that the starry, starry sun was beneath, the starry, starry sky still above, that your ears were down and your toes were pointing upward toward the green, green grass of earth hurtling across the heavens?

Turning up the volume of eye games and wordplay, what became of noon? High noon without the sun above falls as stillborn as a white Christmas without snow. What happened to our perception of free will about the free will of our perceptions? Must we quite abandon visual images of actions on the sun and supplicate ourselves before the necessity dictated by the actions of universal atomic clocks? Is high noon on the sun no easier to find than high noon on the Internet? Is there a limit to how far the analogy will stretch without snapping? Without changing our definition of high noon, requiring the sun to be just as high as high can be, can the skeptic no more visualize noon on the sun or Internet than an all-powerful God can create a stone he cannot lift?

Must the skeptic plead free excuse, claiming the predetermined travel-unworthiness of images and words? Is *noon* an earth-word, not a sun-word, not a cyberspace-word, not a universal? Is the sun in our tale too high, or too lowly? Is pulling the phrase *high noon* out of a hat in one world to explain *high noon* in another but metaphysical sleight of hand? Are skeptics justified in believing that universals are nothing more than stories of a land whose borders have yet to be discovered or explored? Or, are universals merely patterns whose exceptions have yet to be discovered or imagined? Such is the dilemma of trying to wring biomythological explanations from borrowed actions, eye games, and wordplay.

The Skeptic's Lexicon

Words perplex. Like wolves howling at the sight of a full moon or sweater, words are transfigured by context, each context a new world. In the world of gravity, *up* jumps toward the ceiling, *down* falls toward the floor. Not so when a five-year-old enters the world of paper. Now *up* and *down* are defined by the heart; the top and bottom of the letters and paper on the desk are further from and closer to the organ beating inside the child's chest. In the world of gravity, Daddy's leather chair is always the same chair, maintaining its identity independently of the toddler's careening perspective or the chair's orientation. In the paper world, with its ink or graphite upholstery, the chair becomes a *b* or *d* according to the position of the child's ears. With the chair-shaped *b*, the "chair back" is closer to the left ear; with *d*, the "back" is closer to the right.

The skeptic realizes that words often break down between worlds. Just as the words of the paper world do not equal the words of the gravity world, so the words of the latrine do not equal the words of the vanity table—not all toilet water smells as sweet.

The loss of a word's meaning when removed from the context of its founding activity or "form of life" poses problems. In Wittgenstein's words:

> When we do philosophy, we are like savages, primitive people, who hear the way in which civilized people talk, put a false interpretation on it, and then draw the oddest conclusions from this.[242]

Wittgenstein captures how skeptics like me do philosophy, but we still dare to philosophize that as logical as it may be to be certain we have *all* the kids in the car or, after counting, *all* the marbles in the jar, the logic of *all* becomes elusive when applied to *all* the stars in the universe and downright mysterious when language tries to carry us to the *All* beyond the stars.

All, as applied to the kids in the car, or marbles in the jar, is an action word; it belongs to the actions of playing kids or playing marbles. *All*, applied to stars or power, is a story word; it applies to the act of telling, or selling, a story. The story cannot be acted out in other ways. *Dog*, in the sense of "Fido is the dog in the corner," is an action word. We can scratch, feed, or walk that particular dog, to the dog's delight. *Dog* in the sense of "A dog has four legs" is a story word. Such a dog cannot be delighted by a ball or bone. You cannot scratch, feed, or walk it. Story words—the words in the stories we tell ourselves and sometimes others—are not necessarily testable by action. *The planet Pluto* is a good example. Story words are thus perfect candidates for our skepticism.

171

Another question of language arises. Can an all-powerful God create a round square, inside or outside of a story? Rather than question the meaning of *all*, Saint Thomas Aquinas originated a principle that "has been accepted widely among philosophers and theologians. It states that the fact that God does not have the power to do what it is necessarily impossible to do does not undermine his omnipotence."[243] In other words, with God all things are possible except when philosophy tells us they are not. Contemporary philosopher Yujin Nagasawa goes on to explain, "According to this principle, the fact that it is impossible for God, for example, to draw a square circle ... does not threaten His omnipotence."[244] Can an all-powerful chair be oriented like a *d* in the den and a *b* in the bedroom? Such is the risk of equating the words between the worlds of gravity, paper, and transcendence. Such is the risk of trying to find noon on the sun.

What's true of language games is no less true of eye games. Consider the problem of words from another perspective, one with which I am more familiar: vision. Look at this cube:

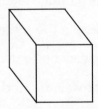

In viewing it, we export how we learned to do seeing in the three-dimensional world (or how our genomes learned to do seeing in their three-dimensional past lives). We misapply these habits to do seeing in the flat world of paper. The cube is a lie. It is not a cube at all. It is not three-dimensional. It is an illusion. It is just a flat square and a couple of flat, lopsided diamonds. The cube is ideal, a mere idea in the mind. The cube does not exist in matter. The only matter involved is ink on a flat page, and even that is imperceptible when viewed from the side.

When we transfer perception from one world to another, it can break down, lie to us, and create illusion. The moon at the horizon appears to change size. From a tall building, people on the sidewalk below are mistaken for ants. We look into the two tubes of the binoculars and see one image. Mountains in the distance are tiny. All lies; all illusions.

If seeing is not necessarily interchangeable between worlds, then are words based on seeing necessarily interchangeable between worlds, or do the pictures and words of action in one world reduce to metaphors in the next? The story of natural selection was first told to aid in the action of predicting the past, whatever that might mean. Applied to values, it became a license for terror. Physicists tell us that an unmeasured electron exists only as a wave of probability, whatever that might mean. Is the word *exists* the same for "The lines *exist* on the paper," "The cube's third dimension *exists* in the mind," and "The unmeasured electron *exists* nowhere in space"? How does God exist—like the line, the cube's third dimension, or the unmeasured electron? Or, is God too busy making the selections for Mother Nature to have time for such nonsense? Is God out of time?

"Time flies like an arrow; fruit flies like a banana."[245] Language is a tightrope act performed above a net of confusion; chaos lurks around the edge of every letter.

The problem has concerned skeptics since at least the ancient Greeks. In Plato's *Phaedrus*, our know-nothing friend Socrates describes how some words allow us to "cut up each kind according to its species along its natural joints, and to try not to splinter any part, as a bad butcher might do." The word *thumb*, for instance, appears to cut nature at the joints. Other words are less precise. As Socrates explains about his previous discussion of the story word *love*, "Whether its definition was or was not correct, at least it allowed speech to proceed clearly and consistently with itself." In other words, words describing real actions may cut nature at the joints; story words need not, so long as they meet the agreements of those telling the stories.

When it comes to our concepts cutting nature at its joints, we apparently suffer from an eye-word coordination problem that dramatically increases when cutting along the joints of a story rather than an action. Take attention deficit disorder with hyperactivity, a diagnosis in which the joints of nature, rather than being cut, were shattered into a half-dozen pieces. We learn in the *Diagnostic and Statistical Manual of Mental Disorders (Third Edition)* that the disorder once enjoyed such sobriquets as hyperactive reaction of childhood, hyperkinetic syndrome, hyperactive child syndrome, minimal brain damage, minimal brain dysfunction, minimal cerebral dysfunction, and minor cerebral dysfunction. This malady of misdirected attention got tossed from the canon of mental illness along with homosexuality. In the next statistical manual, attention deficit hyperactivity disorder (ADHD), one of the more profitable disorders in the history of ontology, joined the ranks of biomythology. In all honesty, I should confess that the terms used to capture the vagaries of perception (my area of optometry) come no closer to cutting nature at its joints. That story I cannot tell, however, without using Ockham's razor to slit my own professional throat.

Dawkinization

Words, as the skeptic intuits, change their definitions to fit the music of the times. To understand the power of musical definitions, we must once more turn to Richard Dawkins. Entomology studies the buzz of bees; etymology, the buzz on *bs*. Evolutionist and Oxford rhetorician Richard Dawkins studies both. Dawkins received his training in zoology and now studies the evolution of words as he might any other species selected by our natural ears, mouths, and minds. In 1976, while reconfiguring terms to fit his title, *The Selfish Gene*, he wrote, "My definition will not be to everyone's taste, but there is no universally agreed definition of a gene. Even if there were, there is nothing sacred about definitions. We can define a word how we like for our own purposes, provided we do so clearly and unambiguously."[246] What could be simpler? While poetic license allows the altering of words to preserve meter and rhyme, musical definitions allow the redefinition of words to preserve theory. Inspired by Dawkins, I use the device throughout this book.

Specious Species

Take the definition of the word *species*. When God said "Let the earth bring forth the living creature after his kind," what exactly did He mean by *kind?* Are all horses of a kind, or only those who have managed to avoid extinction? Are modern classifications from the mind of God or Carl Linnaeus (1707-1778), the father of modern classification? Scientists prefer the word *species* to *kind*, but again, what does *species* mean? According to Ernst Mayr, "arguably the greatest evolutionary theorist since Darwin,"[247] "A species is a group of actually or potentially interbreeding natural populations reproductively isolated from other such populations." Per the definition, one wonders what to do with asexual critters such as bacteria, viruses, molds, worms, yeast, mushrooms, algae, and the like. There undoubtedly is a whole lot of *bringing forth* going on, but where is the *breeding?*

Well, as Dawkins assures us, we can define terms for our purposes. He writes, "Indeed, a species can pretty well be defined as a set of animals that engage in gene transfer among themselves."[248] Like the last amphibian transferring its genes to the first lizard?

In the same paragraph, Dawkins provides "only three tentative exceptions." Apparently, a species is a species except for when it isn't, and Darwin's *On the Origin of the Species* was about—well, we are not really sure, but the more honest title, *On the Origin of the Whatchamacallit,* probably would not have sold out on the first day. Apparently, a universal definition of *species* does not belong to the universe any more than a universal definition of *life* or a universal definition of what it means to be human. In the universe of science, exception is the rule. Looking for absolutes? Try Plato's Forms. Try eternity. Try God. Don't try Mayr or the *Diagnostic and Statistical Manual.*

Whatever their shortcomings, it's a small wonder that musical definitions are so popular. Politicians, salespeople, theoretical scientists—none could function without the ability to say whatever they wish and claim it meant whatever they wanted. As noted in our "Evolutionary Equivocation" chapter, Darwinism can be organismic capitalism or any other form of descent with modification—genetics, epigenetics, evolutionary developmental biology—whatever is necessary for the argument at hand. It appears that musical definitions are here to stay, even if their definitions are not.

Equal Signs Are Not Equal Signs

Outside of math, musical definitions flourish. What, the skeptic may imagine, do equal signs know about equality in the physical world below the ideal world of numbers? When was the last time you saw something in which all the atoms were arranged equally to all the atoms in something else? How could you cut an apple in half and not lose a few quarks so the halves no longer add up to the whole? Similarly, no two apples could equal any other two apples: the number of apples may be equal, but the apples cannot be.[249] Searching for equality in objects is about as fruitful as searching for oral hygiene in amebae. In the real world, equality more likely equals lack of perception.

Numbers are equal, but individuals are not—variations in dental work have been used to prove the inequality of even charred corpses. Reducing the mystery of human transcendence to equal signs misses the orthodontia and the humanity. That engineers worship the equal sign and try to live up to its inspiration merely stresses yet again why Plato was forced to replace the shadows in his cave with a higher world of transcendent Forms. Possibly not since the Garden of Eden have any two fruits hanging on a tree *equaled* one another.

Equal signs apparently suffer from original sin just like the rest of us, but engineers have nevertheless been granted the salvation needed for building buildings, bridges, and flights to the moon. Where musical definitions are concerned, *equal* may mean "the same number" as in math or "close enough for science" as in apples and species. In real science, this amounts to close enough for results. In biomythology, close enough for persuasion may be all that's required, just as in skepticism certain enough for action may suffice, absolute certainty being largely unobtainable.

Rules of Thumb

The problem of words goes beyond those who worship such absolutes as *equal* or *God*. The skeptic should consider a simple concept firmly rooted in the mundane: *thumb*. According to agreement, the word stands for that "short, thick first digit of the human hand, set lower and apart from the other four and opposable to them."[250] The real question, however, is to know whether *thumb* is a real product of nature or simply an agreement, a distinction formed in the minds of anatomists and grave robbers so they can agree on what they are buying and selling. Is *thumb* in your mind or on your hand?

A silly question, perhaps, but if we were to align a ruler with the outer side of the index finger, its straightedge would extend to approximately pass through a crease or line on the palm side of the hand. Palm readers call this crease the line of escape and use it to separate the thumb from the palm. You can easily see this line of escape carved into your own hand and providing escape from worries about anatomy. Even so, the skeptic may reasonably ask if the word *thumb* is carved in stone, flesh, or the mind. Is it carved by action?

Imagine a turf battle between the Union of United Hitchhikers and the World Federation of Palm Readers. Which organization owns the line of escape? Does it belong to the hitchhikers or the palmists, the thumb or the palm? Or, does a boundary run down the middle of the line defining the property rights? If so, who is responsible for drawing the boundary? Might those whose travels depend on thumbs have surveyed differently than those making a living off palms?

Try an experiment. Look at the palm of your right hand. Without moving your wrist, rotate your rigidly straightened right thumb in the widest possible arc. What exactly moves? Anatomy aside, functionally speaking, where do we draw the line ending the movement? When the word is made flesh, where does *thumb* end and *palm* begin? Does the boundary fall on the line of escape or at the lifeline, that arc etched in flesh separating the sensuous "mount of Venus" from the rest of your palm? When you sprain your thumb, what hurts? Where does the brunt of the swelling occur? No matter. Palm readers want the mount of Venus for their own, and anatomists will continue to cite muscle, bone, and ligament to save the cherished definition of *thumb*. The music of definitions transcends the joints of nature. As any good skeptic could tell you, the melody is turf.

The Music of Turf

Each profession hears its own music. Whether you are selling Ritalin, hickory switches, gluten-free food, physical therapy, counseling, occupational therapy, hardcore phonics programs, or vision therapy largely determines whether these shenanigans of attention are characterized as "ADHD," "lack of discipline," "food allergies," "underdeveloped core strength," "sensorimotor integration deficits," "emotional stress," "dyslexia" or "convergence insufficiency." Indeed, the actions used in treating any of these diagnoses might, in a given case, improve attention. Still, do the diagnoses necessarily represent realities in nature any more than the past prejudices of authorities represent those private pictures and words making up the stories known as the mind?

The Infinite Regress

A favorite tool of turf is the use of musical definitions to preserve theories. As history proves to the skeptic, when observations can no longer be stretched to fit theory, words most certainly can be. Consider the infinite regress. When it comes to the creation of our universe, where does the buck stop? In most Western religions, the buck stops with God, the prime mover unmoved. Such reasoning allows us to put a halt to the infinite regress, to silence the four-year-old or philosopher stuck on the question, "Why?" Past the boundary of brute facts, all roads lead to "Because I said so!" And, when that doesn't work, "Because God said so!"

Not everyone listens to God. Richard Dawkins characteristically considers it regressive to use God as a suitable end for a regress. He argues that nothing precludes something from moving God. As he puts it in *The Blind Watchmaker*, "To explain the origin of the DNA/protein machine by invoking a supernatural Designer is to explain precisely nothing, for it leaves unexplained the origin of the Designer."[251] Michael Shermer concurs, denying that God could be the prime mover unmoved behind the Big Bang. His reasoning? *"What* made God, and *how* was God created?"[252] Or, if you prefer, "Who created God?"[253]

Why these protests? The skeptic of naturalism suspects that Dawkins and Shermer are stinging from their own humiliations at failures of a scientific infinite regress. As we will examine,[254] when all anecdotal evidence is excluded—as good science demands—what method is used to prove that the scientific method used for matter can be applied to mind? What method will be used to prove the method used to prove the scientific method? What method will be used to prove the method used to prove the method used to prove the scientific method? And, so it goes—on and on and on. Either creation stops with the prime mover unmoved, and scientific meta-meta-meta-analysis stops with the prime analyzer unanalyzed, or both must pay homage to the infinite regress. Skeptically speaking, plugging an infinite regress with God or with biomythology creates the same problems for the radical skeptic. Both must be taken on faith. As surely as survivors survive, mystery is mysterious.

Cutting the Uncuttable

Since Dawkins and Shermer have already handed creation over to the infinite regress, the skeptic can do no less for a famously blatant example from scientific cosmology. The word *atom* was derived from the Greek *a*, meaning "not," and *tomos*, meaning "to cut." By definition, an atom is "uncuttable." Suppose we have a golden egg, the tale goes. How many times could we cut the egg in half? Providing we had an infinitely sharp knife and an even sharper imagination, we could keep cutting forever. So how does science solve this infinite regress? With their own prime cutter uncuttable: the uncuttable atom. Sure, children and adults alike love the comfort of the fairytale that "everything is made of atoms." To that the skeptic asks, "Are electrons made of atoms?"

Philosophers deny God the right to create square circles or married bachelors. The faith of Dawkins and Shermer comes with no such logical limitations. Cutting the uncuttable, making fundamental particles ever more fundamental and fundamentaler, are not tricks of Lewis Carroll but rhetoric, used to sell the building blocks of the building blocks of the building blocks of the universe. All things are possible to musical definitions, but today the skeptic finds it hard to imagine even the infinite regress of the quark with no strings attached.

Shards of Thought

Yes—words are precarious. If I place my coffee cup on a chair while I sit on the table, can we be certain if *chair* and *table* are better defined by their forms or my actions? Is the table not my chair, the chair my table? Lord Voldemort's nose scarcely resembles Pinocchio's, except that both share the purposes of blowing and smelling. Despite the music of definitions, despite the cuttableness of the uncuttable atom, atomic theory runs submarines and vaporizes cities, but, again, what happens when we come to the joints of the mind?

Now the chaos lurking behind definitions really threatens to tumble us into the abyss. Where exactly does *irritated* become *angry*, *cautious* become *paranoid?* Where does *sanity* become *insane?* Are we classifying thoughts and emotions of human nature according to nature or shattering them, naming the random shards of "faculties" or "modules" according to the whims of the agreements of the time? I can use my eyes to distinguish my doughnut from my coffee cup, but can I see the imbalanced chemicals of depression? Does the imbalance really exist? Is executive function merely "free will with a higher education"? Is the sad beauty of chemical imbalance all in the mind of the beholder? And worse yet, since my mind is invisible to you, just how sweet does toilet water by any other word smell to me?

Trying to answer such questions, the skeptic finds that the musical definitions of consciousness have no chairs in which to sit. When it comes to applying the scientific method to consciousness, the definitions are nothing but agreements about the music. Changing definitions of the mind requires nothing more than changing our minds. There are no joints, only agreements between storytellers. Whether or not we believe there are ghosts in the

184

machine, biomythology is a rhetoric in which there may be no real categories, just errors.

The Problem with Mary

So where does music of definition reside: in the mind, the agreement, or the world? I would argue for all three. Take the word *red*. The actions of physicists and engineers reveals that *red* has certain characteristics related to the physical actions of turning knobs to align needles on dials in the real world. Similarly, you and I can visit an art museum or car lot and point in agreement to which canvases or Caravans are red. Nevertheless, our actions in adjusting our instruments and agreeing where to point our fingers are just part of the story. The other part of the story of the word *red* is told by philosopher Frank Jackson (b. 1943).

The thought experiment revolves around the savant Mary, who knows every word and symbol ever written or yet to be written to describe every physical fact about the color red—the physics, the chemistry, the neurophysiology—everything except the most important thing. Mary has been raised in a black-and-white world. She has never seen the color red. In his thought experiment, Jackson is arguing against physicalism, the philosophy that once we know all the physical facts of the world, we know everything there is to know. For Jackson—at least at the time he wrote his philosophy—the physical facts could never fully capture experience, consciousness, and basic sensations. My point is that when Mary finally experiences the color red, the word has a new meaning to her, a definition whose music once fell outside the reach of Mary's language, much less the reach of biomythology.

Public Words—Private Words

The skeptic may realize that the definitions of words are more than mere agreements; they are inseparable from images and feelings. Take, for instance, the paucity of my own language skills as I try to describe my feelings attached to my mental images of *wife*. All joking aside, I could tell you how upon first seeing the picture of the woman I would one day marry, I laughed with joy for half an hour, how I smiled for months on end, and how I still smile when I recall my good fortune. I might tell you how I traveled to the other side of the world to meet her the day before our engagement party, or how at the airport in Ho Chi Minh City, when we first saw each other, she handed me a bouquet of flowers and hugged me with the gentle affection normally reserved for family members. I might even describe the warmth and softness of her skin as we now hold one another through the night and how the feelings and images do not stop at the warmth or the softness. Per her Asian tradition, she calls me Husband; I call her Wife. Whatever the game of musical definitions produces for *wife*, my feelings revolve around the music, not the agreements of the definitions—those I save for humor. "We had a perfect relationship. Now she speaks English."

So what has this got to do with biomythology? Everything! Instead of *wife*, substitute the word *Darwinism*. Does your *Darwinism* equal my *Darwinism?* As we have seen, the definitions have quite a musical history. Different groups have different agreements about the word. Again, unconnected by the same actions and perceptions, words cannot be connected by equal signs. The definitions of *Darwinism* vary far more than the definitions of *toilet water.* But, this is only the beginning of the problem.

How does Darwinism *feel* to you? Does the word relate one way or the other to images and feelings associated with years of wonder and study and gifted, charismatic, even beloved teachers? Is the word associated with cherished friends, inspiring conversations, dreams for the future, and the comfort of worldviews cushioning us from chaos? Is the same thing not true for creationists and Darwinians? Can they divorce *Darwinism* from the images in their minds any more than I can divorce *wife* from my images of love? When discussing leash laws, can a dog owner divorce the word *dog* from the images of that pet pooch gleefully chasing a ball or gnawing a bone?

When we argue about Darwinism, are we arguing about agreements or about the images and feelings? So long as images, feelings, and past actions exist in our minds, is the objectivity of language not a chimera? Is it not inevitable that musical definitions will continue to be tailored to preserve those images and feelings— those passions, those values? No matter how rationalists rationalize away the passion behind their defense of rationalism, romantics continue to believe that passion transcends language and that language cannot transcend passion. *Democrat, Republican, Christian, atheist, creationist, Darwinian, gay, Obamacare*—such words often cannot be reduced to the agreements they represent, cannot be divorced from the passions they evoke. The skeptic wonders if it safe to judge rednecks by the color of their skin.

Biomythology

Biomythology is the belief that we can ignore the images and feelings that make us human when we are defining what makes us human, that we can ignore what goes on in our minds when we use words and science to capture what goes on in our minds. Biomythology is a meditation in which the mantra is "science" and the images and feelings of words are supposedly freed to float gently into the past. Biomythology is the heartfelt fantasy that language can be reduced to heartfelt definitions rather than the feel of the heart.

But, alas, am I adding straw to my straw man, or am I Dawkinizing my own definition of biomythology to align it with my skepticism about language? Of course I am. We learn from example, so I am teaching by example. Behold the use of musical definitions in the rhetoric of objectivity. The skeptic should watch for them wherever lasting theories are sold.

10

Rock of Evidence

"We live in the Age of Science, in which beliefs are supposed to be grounded in rock-solid evidence and empirical data."
—Michael Sherman, *The Believing Brain*

The skeptic may well ask, "Is 'rock-solid evidence' the imaginary friend of Richard Dawkins?" With the invention of tectonic plates shifting the crust of earth, causing the ocean floor to churn, build into waves, and break into mountains, rocks aren't as solid as the cliché once implied. Still, as every skeptic should realize, evidence in science has a long and illustrious history.

Take the science of burning witches. Before the property of elderly women could be confiscated, rock-hard evidence, data, and scientific testing were required. Witches, bolstered by the devil, were known to float in water. Witches could not produce tears when the Passion of Christ was detailed. Witches had numb spots on their bodies, marks of the devil, to be professionally diagnosed by "witch prickers" adept at shaving bodies and finding insensitive areas. [255]

Sure, the theories behind the evidence have since changed, so the evidence is no longer considered evidence, but in science, this is often the case. The same tear tests I was taught in optometry school in the 1970s (to diagnose dry eyes, not witches) are also no longer considered valid. Evidence, it seems, is in the eyes of the scientist. Here is not the place to go into what evidence some of those eyes have beheld. Readers with an interest may want to begin by Googling "Tuskegee Syphilis Experiment" or "Nazi Science" or "Human Radiation Experiments." Look at the evidence for what it means to explore what it means to be human when what it means to be human is ignored.

Empiricism

The *Shorter Oxford English Dictionary* offers a convenient definition of *empiricism* for our purposes: "Practice based on experiment and observation; formerly, ignorant or unscientific practice, quackery." One might easily, and wrongly, assume that the derivation of *empiricism* comes from "empire," considering how many biomythological empires have been built on it. But no, the derivation is *empeiridos*, "experienced," from *empeiros*, "skilled," literally "tried in," from *peira*, "try."[256] While orthodox medicine practiced and promoted the science of bloodletting, the quacks or empirics used what they had tried and, in their own experience, found to work.

Skeptics have long been suspicious of empiricism. Four centuries before Christ, the skeptic Pyrrho—as reported by Diogenes Laertius[257] seven centuries later—noted the inconsistency of the senses—that apples looked yellow, tasted sweet, and smelled fragrant; that objects reflected by imperfect mirrors changed appearance; that viewed from different distances and perspectives the sun changed size, rough mountains appeared smooth, squares appeared round, straight lines appeared curved. Our perceptions were, at best, questionable. Apparently, nothing has changed. Today, Richard Dawkins, excusing the fact that no one has ever seen an amphibian give birth to a lizard, notes, "Eyewitness testimony, 'actual observation,' 'a datum of experience'—all are, or at least can be, hopelessly unreliable."[258] The skeptic wonders, should we trust rock-solid evidence and empirical data only when they agree with our favorite theory?

Sir Francis Bacon, four hundred years ago and obviously influenced by the writings of Diogenes Laertius, knew as much:

> For however much men may flatter themselves and run into admiration and almost veneration of the human mind, it is quite certain that, just as an uneven mirror alters the rays of things from their proper shape and figure, so the mind, when it is affected by things through the senses, does not faithfully preserve them, but inserts and mingles its own nature with the nature of things as it forms and devises its own notions. [259]

Bacon further cautioned that "it is a very great error to assert that the senses are the measure of things."[260] Luckily, Bacon fixed the fragility of our perception with his method of induction, requiring careful observation of repeating patterns in the real world—strict correlation of, say, perspiration and exercise—before we conclude that sweating causes running. Now, according to Bacon, "with the help of God's goodness, we will have furnished and adorned the bedchamber for the marriage of mind and universe."[261] Thus, the skeptic finds universal scientific connubial bliss, except perhaps when the universe has a headache.

Essentially, in today's science, empiricism demands that we rely on observation— unless observation conflicts with what we demand to see. As historian of science and former Feyerabend student Steven Goldman notes:

> Copernicus's theory requires us to believe contrary to all experience that the Earth is rotating on its axis at approximately 1,000 thousand miles an hour. The Earth's circumference is about 24,000 miles, so in a 24-hour period, we have to cover 24,000 miles. If you throw a ball up in the air, how come it isn't blown backwards? If a bird takes off at the equator, why doesn't a 1,000-mile-an-hour wind ... blow it backwards? Why don't we see any evidence of this motion?[262]

Galileo, confronted with the problems of observation, was forced to discover a new relativistic theory to justify why "heavy bodies ... falling from a height, go by a straight and vertical line to the surface of the earth."[263] Galileo concluded, "It is, therefore, better to put aside the appearance, on which we all agree, and to use the power of reason either to confirm its reality or to reveal its fallacy."[264]

Dispelling what he calls "Galileo hagiography,"[265] Feyerabend writes:

The Church was not ready to change just because somebody had produced vague guesses. It wanted proof—scientific proof in scientific matters. Here it acted no differently from modern scientific institutions ... Hence Galileo was advised to teach Copernicus as a hypothesis; he was forbidden to teach it as a truth.[266]

Having carefully explained how at odds with real experience a spinning, hurtling earth was, Feyerabend concluded:

To sum up: the judgment of the Church experts was scientifically correct and had the right social intention ... It wanted to protect people from being corrupted by a narrow ideology that may work in restricted domains but was incapable of sustaining a harmonious life.[267]

Has tolerance to opposing paradigms grown since Galileo's day? Have humans become less human and more humane? However we answer these questions, scientific empiricism remains the belief that seeing is believing—except for when it is not what science wants us to see or believe. Faith in a weightless God is rejected; faith in a weightless neutrino was embraced for fifty years. Confirmation in physics demands independent, repeatable observations by experts—something that is impossible when studying the feeling of consciousness in others. As the skeptic has already learned, we can demand one standard for electric lighting and another for electroconvulsive therapy most conveniently by dividing *science* into real science and biomythology.

Vision, Stories, Values, and Believing

We are passionate about our seeing—so passionate that seeing is believing. Even so, the skeptic may freely doubt the story that vision is something that happens to us. We may instead view vision as a story we tell. In doing vision, we normally repeat stories passionately learned through past actions in the world.

Suppose, for instance, we view a polished Granny Smith apple before us. We use our eyes to tell the story of a large, round, green, solid, juicy, heavy, crisp, and tart apple just waiting to be devoured. When we explore the apple from the side, however, vision may need to construct another story: an apple cored by ravenous worms. We learn that perception of the apple may include a backside that does not exist, that the perceived whole apple was nothing but a story. Only when trained to sketch the apple from different viewpoints do we learn the value of the story of ever-changing perspective, color, shadow, overlap, and shape inspired by the romance between light and retina.

The skeptical philosopher David Hume cautions: the word *is* springs from objects and their relationships; *ought* springs from sentiments to be found in the breast; reason, in Hume's story, can scarcely bridge the gap between the two worlds.[268] Our visual stories were obviously not penned by Hume. Vision does little else but mix passion, action, and story—does little else but confuse *ought* with *is*.

To understand this remark, the skeptic may turn to James J. Gibson (1904-1979), author of *The Ecological Approach to Visual*

194

Perception. Gibson hypothesized that to perceive the composition and layout of surfaces may depend on perceiving "what they afford." If we consider infant perception from this perspective, then when a baby looks at a breast, he has one question in mind: "What's in it for me?" Not until the baby matures and experiences more (or less) mature actions will vision tell passionate stories of firmness, curvature, cup size, flesh tone, implants, or eroticism.

We see what we value, or have valued, for action. We see what's in it for us. Food, shelter, clothing, lips, beauty, a cliff, a chance to pass an examination to please a professor, or a chance to impress others with how much we *know*—all may afford or provide something; all may have value. Our values determine perception. A hunter has different values than an illustrator; they value different parts of the world in their visual stories. What we include in visual stories is known as figure. What we exclude from our stories is known as ground. Figure-ground relationships revolve around value. The hunter's story includes the prey quartered by the crosshairs of his scope; the illustrator's story includes the foreshortening of the rifle barrel. Thus, whether you are drawn or quartered dictates the story.

To see the link between vision and value for yourself, look about and observe what is before you in the room. Now, take a second look and see all the things you may not have valued enough to notice during your first viewing: the amount of space or air between you and the objects, the amount of space or air between the objects themselves, the shape of the shadows, the variations in colors that first appeared solid, the dust on the floor, the presence of your nose, the shape of your spectacles' frame if you are wearing one. Habitual seeing is a story told according to values set by prior experience and present desires. Visual exploration—visual consciousness, if you will—is the search for a passionate new story about values. Fellow optometrist Larry MacDonald told the story in a different way: "Eyes don't tell people what to see; people tell eyes what to look for."

No visual story can simultaneously capture everything, which is to say that words and pictures interpret, rather than describe reality. Just as seeing is indivisible from the values of the seer, the story is indivisible from the values of the storyteller. The soul of the storyteller is to be found in the soul of the story, not just because of what the teller values enough to tell, but because what the teller does not value enough to mention.

Evidence is Believing

What is true about vision, the skeptic should suspect to be true about evidence. Consider a thought experiment. Imagine that evidence we see is like a book in a bookstore. Suppose you surveyed those leaving the store, asking each what evidence he or she had encountered. Would the evidence rest in the books, or the tastes of the readers? The skeptic finds that those who most shrilly cry, "Give us this day our daily evidence!" are the same ones who most shrilly demand what books we peruse, what pages we turn, what paragraphs we believe. "Pay no attention to the man standing behind the curtain," are the watchwords of empiricism when applied to science.

Not only in bookstores do different observers see different things. When the heavens were examined through Galileo's own telescope by twenty-four professors, their observations hardly confirmed Galileo's claims. Enough of them wrote the astronomer Kepler that he was forced to write to and question Galileo:

> I do not want to hide it from you that quite a few Italians have sent letters to Prague asserting that they could not see those stars [the moons of Jupiter] with your own telescope. I ask myself how it can be that so many deny the phenomenon including those who use a telescope.[269]

Years later, when Galileo had gone blind in one eye, he wrote with the same consummate modesty that had previously endeared him to Church authorities and ultimately precipitated his housebound fate:

> The noblest eye is darkened which nature ever made, an eye so privileged and so gifted with rare qualities that it may with truth be said to have seen more than the eyes of all those who are gone, and to have opened the eyes of those who are to come.[270]

When it comes to blindness and chutzpah, however, Galileo's was nothing compared to that engendered by what has affectionately come to be called "experiment."

David Cook

Experiment

Bacon saw experiment as our salvation from the dictum that "it is a great error to assert that the senses are the measure of things."[271] About our salvation from the *failure* and *distortion* of our senses, Bacon wrote:

> We do this not so much with instruments as with experiments. For the subtlety of experiments is far greater than that of the senses themselves even when assisted by carefully designed instruments ... And therefore we do not rely very much on the immediate and proper perception of the senses, but we bring the matter to the point that the senses judge only of the experiment, the experiment judges of the thing.[272]

Bacon's "carefully designed instruments" are typically those that are carefully designed to show what they are carefully designed to show. In a recent article, "Why Most Published Research Findings Are False," the unfashionably skeptical author argues:

> Simulations show that for most study designs and settings, it is more likely for a research claim to be false than true. Moreover, for many current scientific fields, claimed research findings may often be simply accurate measures of the prevailing bias.[273]

Such bias appears to escalate when "there is greater financial interest and prejudice." Thus, anything connected to a profit or worldview is at once suspected.

198

To Experiment or Not to Experiment?

Despite Bacon's optimism, experiments have been contested ever since the beginning of modern science. Alchemy and its quest to turn lead into gold were still quite popular in the seventeenth century. The subject fascinated Sir Isaac Newton. Similarly, Robert Boyle[274] (1627-1691) was impressed with the American alchemist George Starkey (1628-1665). From alchemy and its first cousin, chemistry, science derived its preoccupation with number, weight, measurement, and experiment. Boyle was part of the group that helped establish experimentation. In 1663, he wrote *Some Considerations Touching the Usefulness of Experimental Natural Philosophy*. He also gave us Boyle's Law about the relationship between volume and pressure of gases. To derive the law that would later carry his name, he devised experiments using an air pump. Boyle relied on the immaterial "spring of air" to explain his findings. Materialist Thomas Hobbs (1588-1679), who in *Leviathan* authored the idea that governments were necessary to prevent man's natural propensity for chaos and a war of all against all (however good governments themselves were adept at chaos and war), similarly saw the human mind as merely matter in motion and was not kindly disposed toward Boyle's immaterial "spring of air."

Hobbes responded to Boyle's air pump experiments, in the words of science historian Frederick Gregory, with "his own treatise *against* experimentation. Hobbes opposed experimental philosophy because, he said, it was not really philosophy. Natural philosophers should explain by identifying the causes of things, not by experiment."[275] About the dispute between the atheist Hobbes and the Christian Boyle, historian Gregory concludes:

> As the debate between Boyle and Hobbes reveals, experimentation in natural science cannot be the straightforward objective procedure it is sometimes portrayed to be. Differences in philosophical or religious outlook affect how an experimental procedure is understood. And many other factors condition the perceived meaning of significance of an experiment, not the least of which is the position in society of the person evaluating it.[276]

While this may not be true for Shermer's "rock-solid" version of experiment, it apparently applies to softer versions, like those supporting Boyle's Law.

Experiment Defined

Experiment is the art of quoting nature out of context. At best, experiment reduces reality to analogy—predictably false from some perspective. Experiment sells out nature to sell theory. Universals dissolve past the edge of the experimental universe. Dropping large and small ball bearings in a vacuum hardly predicts the results of dropping a balloon and a bowling ball off your bedroom balcony and onto the head of your neighbor's howling cat.

Eva Jablonski and Marion Lamb are postmodern evolutionists. They are skeptical of the outmoded, mid-twentieth-century *modern synthesis* version of Darwinism (the one taught as close-enough-for-science in public high schools). They write about the experiments of the Augustinian friar Gregor Mendel (1822–1884), who is a popular candidate for the father of the modern science of genetics, "A crucial part of Mendel's findings was that, with the strains he chose to use (and he made his choice very carefully), hybrid offspring did not show intermediate characteristics; they resembled one or other of the parents."[277]

Had Mendel used many varieties of plants for his experiments, his results would have been quite different, even equivocal. However rarely life conforms to the simplistic predictions of Mendel, we nevertheless continue to make the analogy that his experiments are good analogies for life in general and that there is a "gene" to be found for ADHD, dyslexia, and your short game in golf.

As Jablonski and Lamb write about single-gene diseases such as sickle-cell anemia:

> Such simple monogenic diseases are not common; they make up less than 2 percent of all the diseases that are known to have a genetic component. For the remaining 98 percent of 'genetic' disorders, the presence or absence of the disease and its severity are influenced by many genes and by the conditions in which a person develops and lives.[278]

Compare the world to Mendel's experiment and you'll be wrong ninety-eight times out of a hundred. Despite such complexities, we continue to be plagued with what Jablonski calls "genetic astrology," the idea that biomythologists will be able to look into our genes and predict our futures, including our behaviors and our proneness for disease. About applying such science fiction to behavior, Jablonski and Lamb write:

> With complex traits like mental ones, tens of hundreds of genes ... are involved in the construction of the trait, and we do not even know how to define 'the environment.' There are so many social and psychological environments that may be relevant to development—almost as many as the number of people. And these environments are partially constructed by the behavior of the individuals themselves![279]

Before Our Very Eyes

Not all experiments are created equal. The better ones can be repeated and the same results obtained, which is to say the experiments can be "reproduced." Darwinians, handicapped by the unfathomable depth of time, can offer no repeatable experiments on the forks in the tree of life. To solve this problem, Richard Dawkins cites the work of Richard Lenski and colleagues, who have raised 45,000 generations of *E. coli,* a common bacterium. Using artificial environments (thus essentially studying artificial, not natural, selection) Lenski has trained his *E. coli* to sit, roll over, play dead—even to eat citrate rather than glucose. Despite the artificial environment, despite the organisms being asexual so the selection—natural or artificial—can give them no sexual advantage and the traditional definition of "species"—as those able to breed together—does not apply, Dawkins cites the study as an example of evolution occurring, according to his chapter title, "Before Our Very Eyes."[280]

Despite playing the game of evolution for 45,000 generations, none of the *E. coli* has become a streptococci or cockroach any more than century after century of breeding dogs has turned them into cats. The study really says nothing about how any major fork in the evolutionary tree occurred. Sure, finches' beaks and moths' dress coats may oscillate to the whims of environmental change and genetic expression, but Darwin's tiny biomechanical variations tuning the forks in the tree of life, as Darwin imagined and high-school biology classes are taught, may still live on only as a fairy tale. The bright side of the experiment is that Lenski did not introduce the *E. coli* to God or religion, so at least Dawkins, Sam Harris, and Daniel Dennett can't condemn him for "child abuse."

Experiment and Perception

It is easier to perform a physics experiment than a psychological experiment in a vacuum; the subjects die. Experiments of the mind create illusions in oxygen-deprived subjects and researchers alike. To see what I mean, let's try an experiment of the mind. We'll call it "The Floating Finger Experiment." (If you have a lazy eye, the experiment will probably not work, but Galileo's telescope didn't work for everyone either.)

> 1) Raise both of your hands to eye level about eight or ten inches in front of your nose. With your fists otherwise closed, extend your pointer fingers, left

pointed right, right pointed left, until their tips touch at the exact level of your eyes. Be sure both pointer fingers are parallel to the ground and parallel to your face.

2) Look at the tips of the two pointer fingers and see they are touching, exactly as you would expect. Then, without moving your head or the fingers, look "through" the fingers as if you were looking straight across the room. (Some may have to lower the fingers by a half an inch and view the distance over the top of the fingers.)

3) A tiny "finger" with a nail at each end should appear between the two real fingers. If no such third "finger" appears, keep looking all the way across the room and wiggle your pointer fingers, still touching, until the tiny finger appears.

4) While still holding your gaze across the room, move your two pointer fingers about a quarter of an inch apart; the tiny finger should appear to float between the two pointer fingers.

5) Return your gaze up close. Look directly at your fingers; they will once more appear in their correct location and the tiny finger will disappear.

Most of us excuse the floating finger as an optical illusion. Illusion is the product of taking perception out of context, a product of taking nature out of context as experiments take nature out of context. We perceive illusions when we apply the seeing habits learned in old worlds to new worlds; for instance, we imagine 2-D pictures and movies are 3-D.

In the words of philosopher Alva Noë, "Perception is not something that happens to us, or in us. It is something we do."[281]

A brain-injured person can often no longer *do* perception. When moving our bodies in an abnormal way (pointing eyes far while attending close) we can no longer *do* perception. When acting in a rearranged or "new" world (a different finger in front of each eye), we can no longer *do* veridical perception.

Doing perception requires habits we have developed between our brains, body movements, and the world. To break those habits is to break perception and action. You have learned how not to spill your milk while reaching for your glass. Distort your brain with alcohol, tie a weight on your wrist to disturb your normal movement patterns, or change the lighting and reduce the size of the glass to pervert depth perception, and you may soon be crying over spilt milk.

To consider how experiment is the art of taking nature out of context, consider the Nobel Prize-winning work of Hubel and Wiesel, who ingeniously showed how stimulation of cells in the back of the eye (retina) may be linked to stimulation of cells in the back of the brain (visual cortex). The two researchers explained their method:

> Recordings were made from forty acutely prepared cats, anaesthetized with thiopental sodium, and maintained in light sleep with additional doses by observing the electrocorticogram. Animals were paralyzed with succinylcholine to stabilize the eyes. Pupils were dilated with atropine.[282]

In the Italian Middle Ages, the poison atropine was used to dilate the eyes of beautiful women, the *belladonnas*. The treatment not only gave the women seductive large, dark eyes; it perverted their vision within arm's reach, blessedly soft-focusing the ugly men who embraced them. That the Hubel and Wiesel cats were sexy is beyond doubt. If, however, perception includes brain, movement, and environment, and changing any one of the three can create illusion, what, if anything, does the study of sleeping, paralyzed, sexy cats tell us about vision perception? Such a leap from neuroscience to perception is a leap of faith. Clinicians nevertheless used the findings to justify (wrongly) the still-popular clinical myths denying older children and adults treatment for their lazy eyes (amblyopia), myths that were contradicted, but hardly eliminated, by later clinical trials.[283]

Again, illusion is the result of *doing* perception in a new world (environment) as it was learned to be done in an old world. Any psychological or neuroscience experiment that alters the brain with surgery, restricts eye movements, or changes the normal patterns in the world takes the chance of capturing illusion better than it captures what it means to be human. This is why real science has given us light switches and vaccines for polio, while biomythology has given us the *Diagnostic Statistical Manual* requiring revisions as homosexuality escapes the closet of mental disorders.

The Repeatablility of Miracles, Magic, and Mind Experiments

One recent study[284] estimates that 50 percent of life-science pre-clinical research—$28 billion's worth a year—could not be reproduced, thus questioning the validity of the experiments. One wonders what percentage of the studies studied centered on mind rather than matter. To decide if a drug kills bugs falls in the province of real science. To decide if a drug increases comfort or mood falls in the province of consciousness and thus the province of biomythology, in which, to achieve desired outcomes, failed studies are not published, and experimental subjects are selected as carefully as parents select friends for their children.

Another recent study, not surprisingly, tells us that in the behavioral sciences almost two thirds of a hundred studies that once passed as truth in reputable journals similarly failed the reproducibility criterion. Andrew Ferguson, a senior editor of *The Weekly Standard,* in his article "Making It All Up," tells the story of how such studies are nonetheless promulgated to tell us what it means to be human. As an example, he quotes from *New Yorker* staff writer Malcolm Gladwell's bestseller *Blink*: "[Such experiments] suggest that what we think of as free will is largely an illusion." We will take a skeptical look at the free-will-versus-free-excuse debate in a later chapter,[285] but in Gladwell's defense, his quoted material actually said nothing one way or the other about free will. Gladwell's use of the word *suggest* is a caveat confessing that his interpretation of the experiments could be as false as he believes

our illusions on free will to be. Ferguson is less charitable about

Gladwell, but provides a more profitable outlook for the skeptic:

> *Blink* sold more than two million copies. Gladwell
> became a sage to the nation's wealthiest and most
> powerful businessmen and policy makers, bringing
> them the latest word from the psych labs—a hipster
> version of the ancient soothsayers placing chicken
> entrails before the emperor.[286]

Ferguson's article essentially cautions the skeptic to be

diligent in detecting what our polemic has identified as biomythology.

As every skeptic should suspect, we are indeed no more perceptually

equipped for mind experiments than we are for magic tricks and

miracles. Applying old seeing habits to new situations, as we have

seen, invites perception (and action) to break down. As we do

perception in our old way, we may quite possibly create illusion or

perceptual blindness. Subjects of psychological experiments are

routinely lied to about the purpose of the experiment. Acting in an

illusory world, subjects are tricked into providing researchers with

illusionary data confirming the researcher's illusions about what it

means to be human.

Witnesses of crimes are similarly fallible; most are not

practiced in witnessing murders and so seldom agree when asked,

"What happened?" If news reports are to be trusted, dozens of

bystanders once stood numbly by while a woman on a New York

street was attacked and cried for help. Did the bystanders lack altruism? More likely they lacked the practiced perception and behavior needed to act in the novel circumstance; they had not attended hero boot camp. Like children reared indoors and slow to play ball outside, they stood and watched—dumbstruck and numb.

The same problem arises during miracles. How likely would we be to perceive a miracle even if it occurred? If a dead man who was lying beside the road were raised, would anyone notice, or would one edit the miracle to fit his or her expectations? (Why did the dude in the robe wake the guy sleeping beside the road?)

In our floating finger experiment, how could we tell if the two-nailed finger hanging in the air was a miracle or a mind experiment? Similarly, magic tricks work because we apply our old way of seeing to the new world of magic. The magician, prepared by practice, sees the magic for the illusion it is.

What might the good skeptic make of all this? So long as mind experiments are dignified as "science," it is equally scientific to believe in muggings, magic, and miracles.

Again, the only lesson to be learned in experiments regarding perception or consciousness is that if we surgically alter the brain, restrict eye movements, or introduce a novel environment, perception breaks down. Understanding the mind is an art, not a science, and artists are masters of deception. As mentioned earlier, we sit in the movie theater watching lights flash on a flat screen and imagine both depth and movement where none exists. Perception is something we do, and we need to learn to do it differently if we are not to be deceived in novel environments. Allowing mind experiments to tell us what it means to be human is like motion pictures telling us that the images on the screen are three-dimensional and moving. When it comes to perceptions derived by experiment or technology, the skeptic must be able to willingly suspend or not suspend disbelief.

It is doubtful that many such experiments even qualify as science. In the words of Robert Hazen, a Carnegie Institution research scientist with 220 articles and fifteen books to his credit, "Science is a way of knowing about the natural world, based on reproducible observation and experiments."[287] The skeptic understands that until we can see the world through each other's eyes, until I can see what the color blue looks like to you and you can see what the color blue looks like to me, then experiments claiming to capture conscious experience hardly fall within the realm of science. We have no way of knowing what we are trying to repeat.

Fact-Dropping

Another euphemism for empiricism is *fact-dropping*. The skeptic should be aware that whatever Darwin's talent for dropping names, it was nothing compared to his talent for dropping facts. Paul Johnson writes:

> He saw himself collecting material for his great book so as to present his theory backed by overwhelming evidence, and it is certainly true that he amassed an ever growing pile of facts about the way organisms

grew. As he put it, 'I am like Croesus, overwhelmed by my riches of facts.' The term overwhelmed was apt. Like many other scholars of all times, Darwin accumulated more material than he could ever possibly have needed.[288]

Were we talking about mere description, this observation would be correct. Having written biographies on Socrates, Jesus, and Churchill, Johnson has undoubtedly learned through experience the temptation of accumulating too much material. In addition, Darwin may have been overwhelmed by the enormity of his task. However, a primary rhetorical technique of that task was overwhelming his audience. Darwin was not merely telling a story; he was making a sale.

Origin is replete with particulars in every dimension and direction of the natural world, leaving no doubt of Darwin's command of his subject. Yet, there are important differences between dropping names and facts. While it is permissible to drop the name of defeated politicians—"I used to play golf with President Ford"—it is less effective to drop the name of politically defeated facts—"I used to play golf on the planet Pluto." There is, however, a far more important distinction: the dropped names should be famous and the dropped facts obscure.

Darwin was a master at the technique. In his "Difficulties on Theory" chapter, he explains how natural selection could have created the eagle's eye from a paramecium's ingrown toenail:

In certain crustaceans, for instance, there is a double cornea, the inner one divided into facets, within each of which there is a lens-shaped swelling. In other crustaceans the transparent cones which are coated by pigment, and which properly act only by excluding lateral pencils of light, are convex at their upper ends and must act by convergence; and at their lower ends there seems to be an imperfect vitreous substance. With these facts, here far too briefly and imperfectly given, which show that there is much graduated diversity in the eyes of crustaceans ... I can see no very great difficulty ... in believing that natural selection has converted the simple apparatus ... into an optical instrument as perfect as is possessed by any member of the great Articulate class.[289]

Crustaceans, cornea, facets, lens, cones, convex, vitreous, Articulate class—if the length of the sentences is not enough to take your breath away, then the onslaught of facts and technical nomenclature certainly will. Had Darwin written for trained naturalists with backgrounds in taxonomy and ocular anatomy, the above *might* have qualified as plain prose. Since Darwin was writing to convince the world, however, his complexity tips into rhetoric.

Selling his theory by attaching it to his extensive accumulation of exotic facts about nature is much like a classical scholar selling a theory by quoting passages written in Latin or Greek. Admiring the scholarship, we are more likely to consider the theory. As we have mentioned before, Rutherford divided science into "physics and stamp collecting." Darwin so impresses us with his stamp collection that we half believe his theory is physics.

Cat in the Hat Empiricism

The stamp-collecting analogy would suffice, but another metaphor deserves the skeptic's attention. In Dr. Seuss's *The Cat in the Hat,* there is an amusing illustration. We find the Cat balancing on a ball and supporting cup and cake and toy ship and books and umbrella and fishbowl and fish and dish of milk bottles. "Cat in the Hat empiricism" is my description for Darwin's rhetorical device of using mind to connect the dots of fact to illustrate a theory. His disciples ape him. Consider Jerry Coyne's fleshing out Darwin's example of eyes and natural selection:

> A possible sequence of such changes begins with simple eyespots made of light-sensitive pigment, as seen in flatworms. The skin then folds in, forming a cup that protects the eyespot and allows it to better localize the light source. Limpets have eyes like this. In the chambered nautilus, we see a further narrowing of the cup's opening to produce an improved image, and in ragworms the cup is capped by a transparent cover to protect the opening. In abalones, part of the fluid in the eye has coagulated to form a lens, which helps focus light, and in many species, such as mammals, nearby muscles have been co-opted to move the lens and vary its focus. The evolution of a retina, optic nerve, and so on follows by natural selection.[290]

Unless mammals evolved from abalones, which evolved from ragworms, which evolved from the chambered nautilus, which evolved from limpets, which evolved from flatworms, this passage is entirely rhetorical. If each dot is a fact—however exotic—the lines connecting the dots are chosen not from nature but theory; the resulting picture would qualify as, at best, abstract art.

Returning to Dr. Seuss, we find the Cat still balancing cup, cake, ship, books, umbrella, fishbowl, fish, dish, milk bottles—items related no more closely than abalones and flat worms. If we connect the facts, is the Cat's balancing act an example of natural selection?

Cat in the Hat empiricism is the bedrock of biomythological rhetoric. The skeptic should realize that adding Piltdown man to Lucy to a chambered nautilus to an umbrella and fishbowl cannot build a bridge or a rocket ship. It cannot light a room at the flick of a switch. It can only convince. A single discovery can slay ten thousand footnotes; ten thousand facts cannot transform a single falsehood into truth, no matter how rock-hard your evidence, no matter how ingeniously empirical the imagined lines between your data.

11

Fallacy to a Higher Power

When do fallacies turn sound reason into the noise of biomythology? Before burning another straw man in effigy for the edification of aspiring skeptics, let's begin with a caveat.

Against Reason

I am a misologist. Just as *misanthropy* is hatred of mankind, and *misogyny* includes hatred of masseuses, *misology* is hatred of reason. My own diluted brand of misology falls far short of hatred; I merely lack a proper reverence for reason, especially when cruelty is allowed to have it reasons. Then the skeptic must, if necessary, be equipped to doubt reason itself. Which is easy enough. While hardly denying that reason may inadvertently stumble upon the truth, the misologist never imagines that because something is more reasonable, it is closer to the truth. He can offer the skeptic any number of reasons for his doubt. Here are a few of my favorites.

Thomas Aquinas (1225–1274), in his commentary on Aristotle's discussion about the intellect, reasoned that "reason proceeds from what is known to inquire into the unknown." If such reasoning is true, nobody who already knows what he is talking about bothers to reason—that is, unless he is persuading rather than exploring. Deciding on a brand of gum may include a reason, but chewing the gum exercises the jaws, not the intellect. The need for reasoning signals to the skeptic that observation or action is failing, that the evidence is not speaking for itself. The use of reason signals that the user either is ignorant or hopes that we are.

Another wonderful analogy for the frailty of reason is provided by the process of confabulation—the invention of imaginary stories to fill gaps in memory. Too often, reasoning mimics that confabulatory process. We imagine logical stories to fill the missing links in observation, and then we imagine logical stories to fill the gaps in the imagined stories. This filling of reality with imagination continues to repeat itself until it comes to resemble a line dance in a house of mirrors.

Fortunetelling provides yet another apt analogy for reasoning. We use imagination to create a series of "reasonable" remarks to predict or direct the future. Just as the fortune-teller's client remembers what fits and forgets what doesn't, we similarly forget the many times our reason was wrong and remember only those remarks which raise the eyebrows and heart rates of our most passionately believed theories and actions. Our reason resembles that of a man fortuitously falling through a manhole to escape a labyrinth; how soon he forgets the thousand perfectly logical dead-end turns that had previously confined him.

For much the same reason, the success rate of reason can be disappointing. Few have illuminated the world like Edison, and yet even he admitted to a thousand failures of reason before inventing a single light bulb. Edison taught by example the keystone of sound reason: keep acting out your fantasies, but do not act out the same mistaken fantasy twice.

Reason doesn't get much better than that.

Still, if Edison's reason was correct only one time in thousands, what chance has my reasoning against reason? Either my reason is false and reason is true, or my reason is true and reason is false. Despairing at the contradictions, I can, in good faith, promise nothing better than rhetoric. Even when reasoning aligns like the sun, moon, and earth during an eclipse, a shift in perspective may turn reason, however sound, into noise.

Another perspective on the limits of reason is inspired by the cognitive scientists Douglas Hofstadter and Emmanuel Sander in their fascinating book, *Surfaces and Essences: Analogy as the Fuel and Fire of Thinking*. The book's title tells the book's story. Its 500 pages support its premise with a stream of examples that puts to shame even Darwin's fact-dropping in *The Origin of the Species*. Thoughts, words, concepts, categories, phrases, actions, abstractions, theories, reasoning—all are analogies of sorts. Take, "What worked before will work again." Or, "She looks like my sister—wears the same earrings and everything." Or, "The force that tugs the apple to the ground is like the force that holds the planets in orbit." Or, "Your pet reminds me of my pet. They both are brindled, have ears and tails, and belong to the canine species."

However, if analogy is truly the fuel and fire of reason, watch out for the conflagration.

All analogies are false from some perspective; otherwise they would be identities. Newtonian physics, for instance, is an analogy that works well for building rocket ships but is false for atomic particles and very fast space flights. If, as Hofstadter and Sander suggest, all reason is based on analogy, then all reason is false from some perspective. Luckily, Newton was not sitting in a black hole, or we might have a theory of applesauce rather than gravity.

Obviously, my reasoning against reason is false from most perspectives, or, as any rationalist could share, my argument is not only false—it is also ridiculous. Despite its failings, reason is the best tool we have for unraveling the world, its mysteries, and myths. Human reason is what separates us from the beasts, allowing us to kill for sport rather than just food and territory. Cruelty is never without a reason. The Crusades, Spanish Inquisition, and pogroms were simply examples of tough love. Right? How but through reason could Augustine have added together "turn the other cheek" and "they that take the sword shall perish with the sword" and found the sum to equal *just war?* How but through reason could Lincoln have traded the evils of slavery for the freedom of death by conscription? More importantly, as we learned during our discussion of the Darwinian religion, how but through reason can we feel awe for this year's version of the theory of evolution?

Reason, it has been reasoned, is what makes us human, allowing us to seek knowledge of self, transcendence, and atomic warfare. Reason is self-evident truth, like the shortest distance between two points being a straight line—except for when it isn't. Unhappily, reasoning itself must be justified by faith. The rules of reason can no more prove the rules of reason than the rules of science can prove the rules of science, or the rules of skepticism can prove skepticism, or the rules of golf can prove the rules of golf. Plato must have guessed as much when he deferred to the transcendence of the higher reason in the great outdoors beyond the cave. Certain faith is common; certain reason, rare. Any martyr can have certain faith, but every fresh discovery or perspective feeds a battalion of certain reasons to the lions.

Our best justifications for reasoning soon descend into rationalization:

"It works!"

"It has changed the world!"

"Without it, chaos would break loose."

"Without it, passion would have no slaves."

Ironically, such rationalizations reduce the truth of reasoning to pragmatism, the same justification I used to define real science. Such justification of reason crumbles into sophistry, which rests on faith in successful outcomes rather than on faith in universal truths.

Reason is especially dangerous when it is used, as I have used it, for rhetoric rather than exploration. Using reason to climb out of a strange bed in the dark and ford a sea of wine-dark lingerie is one thing; using reason to convince a stranger to climb into a bed and swim in a sea of wine-sodden kisses is another. One kind of reason tames new worlds; the other seduces them. One provides salvation past the edge of habit; the other fuels persuasion. The skeptic may assume that, except for conscious exploration and creativity, reason is more often habitual, practiced, and largely unconscious rhetoric riding on the coattails of logic.

Some Reasons Are More Equal than Others

To clarify and build upon the last paragraph, it may be helpful if we divide reason into a couple of definitions borrowed for our purposes from the many alternatives provided by Oxford: 1) The mental faculty, which is used in adapting thought or action to some end; 2) A statement used as an argument to prove or disprove some assertion or belief.

To confuse *reason* meaning "adapt action" and *reason* meaning "prove some belief" is equivocation. One reason provides for negotiating novel circumstances; the other, for fabricating novel propaganda. One revolves around exploration; the other, around sales. No matter how well sales-type reasoning rides on the coattails of exploration-type reasoning, the two types of reason have little in common but the name *reason* and a demand that the parts of the reasoning logically agree.

Suppose, for example, my wife and I always walk from Times Square to Saks Fifth Avenue by taking 42nd Street to 5th Avenue; today, however, we find ourselves blocked by construction on 42nd. We use reason to select an alternate route to Saks. As it turns out, my wife gives me some excellent reasons for detouring through the diamond district. No matter how excellent her logic, the reason she uses to persuade me is not necessarily the same reason she used to plan our route (or justify our route after it had been planned).

Lest the skeptic confuse our two types of reason, I will make a suggestion. Just as I have defined *real science* as the science used to conquer the physical world and *biomythology* as the science used to conquer our minds, so I will define *real reason* as the reason used to adapt personal thought or action to novel circumstances and *rhetorical reason* as the reason used to persuade the thoughts or actions of others. Real reason fuels the discoveries of science. Rhetorical reason sells the politics of science. Rhetorical reason is the universal solvent: it dissolves all arguments and observations conflicting with passion. As we noted in our introduction, what I call *rhetorical reason* philosopher of science Paul Feyerabend has more eloquently dubbed "brainwashing by argument."[291]

Reason—the Slave of Passion

In the extraordinary confession of eighteenth-century philosopher David Hume, "Reason is, and ought only to be the slave of the passions, and can never pretend to any other office than to serve and obey them."[292] An excellent example would be six Granny Smith apples divided equally between two persons, allowing three apples each. The math is as reasonable as reasonable can be, but it is nevertheless a slave to a passion for apples rather than alcohol. Similarly, math may be distilled reason, but if our passion is for distilled spirits—for applejack rather than applesauce—passion calls the shots. Reason merely provides the math for dividing them among the drunks at the bar. This is the reason we instinctively consider the passions, and not just the logic, of the logician.

Common sense dictates that the truth and consequences of the reasoning depend on assessing the passions of the masters, not the logic of the slaves. Such thinking is, however, barred to debaters as the "to the man" or *ad hominem* fallacy. As the skeptic will shortly see, fallacies are the invention of rationalists to protect reason from the ravages of eloquence or common sense. Fallacies, however, are encouraged in the hands of biomythologists.

Valid Falsity and Fallacious Truth

Plato looked toward higher truth. Aristotle looked toward truth's pale shadow—validity. Just as groups confuse agreement with truth, so validity attempts to replace truth with agreement. Aristotle specialized in three-part arguments—proposition, proposition, and conclusion—known as syllogisms. Here are three examples: 1) "All souls are immortal; Socrates is a soul; therefore, Socrates is immortal." 2) "All men are apes; all apes are mortal; therefore, all men are mortal." 3) "All men are animals; all Shirley's dates were men; therefore, all Shirley's dates were animals." Each of the three threesomes is valid, its parts faithfully agreeing. Were the propositions true, validity would demand their conclusions to be true. So, which syllogism is true? Are men souls, apes, or animals? Immortal, mortal, or immoral? Validity could not care less about truth. Validity cares only about agreement. Agreements between lies, between truths, between truths and lies—to validity it makes no difference.

If the parts disagree, the argument is called a "fallacy," the reasoning "fallacious." "All men are fools; Darwin is a man; therefore, Darwin is a fool." This is valid. "All men are fools; Darwin is a fool; therefore, Darwin is a man." This is fallacious; he could be a fool woman. The truth is a different matter entirely. Darwin was a man but no fool, and yet in the above syllogisms his manhood was fallacious, his foolishness valid. Thus, validity can be false, and fallacy true.

When premises are false, validity is like a perfectly structured kiss with no passion, a perfectly structured sentence with no words. At times, it happens that when a premise is false, validity is still strong enough to strain truth out of nonsense: 1) All men are apes; all apes are male; therefore, all men are male. 2) All circles are squares; all squares are round: therefore, all circles are round. Even though all four premises are gobbledygook, both syllogisms are valid. Both conclusions are true by definition. Generally, however, when a premise is false, then fallacy becomes our best hope for truth.

Logical Fallacy

Fallacy is the lifeblood of rhetoric, humor, experimental science, and maybe even love. In a sound argument the logic is valid, the premises true, the conclusion infallible. Fallacious arguments are not sound, but noise. Because noise carries further, it is often harder to ignore. Throughout this polemic, I have introduced skeptics to the noise I have learned from studying Darwinian apologists. We won't bother to count the fallacies in that statement; the Latin might well overwhelm us. Instead, we will move on to a thought experiment.

Mr. Samaritan, Little Miss Logic, and Mr. Fallacy

The following tale highlights some of my favorite fallacies,[293] italicizing them for easy identification. Evaluate the following interchange in which an argument is presented by Mr. Samaritan and the merits of the argument debated by Little Miss Logic and Mr. Fallacy.

> **Mr. Samaritan:** It is raining out, harder by the moment. Your pink dress is getting ruined. The weatherman tells us that there is a 90 percent chance of hail and lightening within the hour. Water is overflowing the curbs. The sidewalks are muddy. It's a long way to even the nearest school. I have opened the passenger door for your convenience. There is more than enough room in my car. Hop in. Let me give you a lift.

Little Miss Logic: Thanks, sir. From the looks of this weather and the increasing frequency of thunder, I would say your argument is sound.

Mr. Fallacy: Be careful, honey. I couldn't help overhearing as I walked by. This guy could be a child molester.

Little Miss Logic: His argument must stand or fall on its own merit. His personal character has no bearing. Don't lay that *ad hominem* hocus-pocus on me. While you are out here poisoning the well, I'd rather be inside the car.

Mr. Fallacy: Everyone knows it is dangerous to accept a ride from a stranger.

Little Miss Logic: No way am I falling for your *appeal to the majority*. Everyone may know, but everyone could be wrong. I'm getting out of the rain, if you don't mind.

Mr. Fallacy: Can't you see the guy isn't wearing any pants under his trench coat?

Little Miss Logic: Spare me the *hasty generalization*. His attire hardly qualifies him as a pervert. You are jumping to conclusions. Maybe he was in a hurry while dressing.

Mr. Fallacy: You are just a precious and innocent child; there is more to life than you understand. I beg you, don't get into the car.

Little Miss Logic: Mixing your fallacies, are you? All that begging and precious-innocent-little-girl sentimentality is an *appeal to emotion*. Your *appeal to ignorance* is no argument for there being a serial killer hiding behind every steering wheel.

Mr. Fallacy: What if I told you that the last child he picked up was found strangled?

Little Miss Logic: Excuse my Latin, but I would say *post hoc ergo propter hoc*. That one event happened after the other proves nothing; correlation does not necessarily imply cause and effect. You are offering superstition, not data.

Mr. Fallacy: He is arguing for the right to kidnap you.

Little Miss Logic: Nice *straw man* fallacy, sir. Mr. Samaritan said nothing about kidnapping. You are putting those words into his mouth, creating a straw man so you can debate against the straw of your fantasies rather than the bricks of his real argument.

Mr. Fallacy: I warn you. Stay out of that car or risk your life.

Little Miss Logic: *False alternative!* Other outcomes are possible. Outside, I may get hit by lightning. Inside, he may take me shopping.

By tossing around predictable fallacies, Mr. Fallacy loses the debate. Little Miss Logic hops into the car to escape the wind and the rain. She and Mr. Samaritan drive away. The next day her body is found in a drainage ditch.

The moral for the skeptic? Failing to persuade can be a greater evil than fallacy. Pure reason, removed from the context of fallacious existence, is not truth. A rabbit can logically rest in a hat, but not knowing you are watching a magic act, you may prematurely give up searching for the trick, assuming instead a coincidence. Assessing logic free of context is risky. The good skeptic realizes that the proof of the rhetoric is in the persuasion. One debater's fallacy is another's best argument. As we saw earlier, fallacies are always invalid, even when they tell the truth.

Suppose we hear the maxim, "Morality is the lubricant of social intercourse. It builds the trust needed for the union of souls." Valid or fallacious? Who said that? George Washington? John Adams? Thomas Jefferson? Would your evaluation change if I told you the speaker was Bill Clinton?[294] Would the words *lubricant, intercourse,* and *union* take on new meanings? Would you wonder if the former President was still training interns?

This argument is, as any good skeptic should be able to see, fallacious for any number of reasons. If you smiled, I was appealing to your emotions. Clinton's name is its own *ad hominem* argument; it poisons its own well with no need for help. "Are you still training interns?" is a *complex question*, which like "Are you still beating your wife?" has no good answer. Also, I used the fallacy of *hasty generalization* by citing a single moral lapse and generalizing it into a pattern of life. I might as well have pointed out a single fish's wetness and generalized wetness into a condition of fish existence without providing further slips or slimes.

Appeal to ignorance, appeal to emotion, appeal to humor, appeal to majority, appeal to authority, appeal to scientific consensus or interpretation rather than data, appeal to aphorism (Experiment is the art of arranging observations to fit theory!), appeal to validity (So what if the truth died? The argument was valid, the operation a success!), appeal to common sense (I saw him pull the rabbit out of the hat!)—all such fallacies are appealing, providing they win the argument and save the life of the child.

If you are interested in the many fallacies applied by Darwin's apologists, peruse Jason Lisle's primer on logic for arming fundamentalist homeschoolers and anti-evolutionist debaters, *Discerning Truth: Exposing Errors in Evolutionary Arguments*. Lisle describes the fallacies used to sell evolution.[295] *Reification*, for instance, is attributing a concrete characteristic to an abstraction, such as Mother Nature pruning her garden, natural selection scrutinizing life, or selfish genes washing and waxing their survival machines. My absolute favorite fallacy, however, is one that forms the bedrock of not only Darwinism but also the scientific method.

The Cats and Dogs of Affirming the Consequent

The skeptic should be aware that some fallacies are more equal than others. Consider an example: If it were raining cats and dogs, then the sidewalk would be wet; the sidewalk is wet; therefore, it must be raining cats and dogs. The logic is false, invalid, fallacious, bogus—take your pick. I could have just watered the lawn. All those dogs and cats could have peed on the sidewalk. Water balloons, garden hoses, spilled drinks, incontinence—the ways in which the logic could fail are bounded only by imagination and discovery. Consider a second fallacious example: If it were raining cats and dogs, then I would not be wearing a sundress; I am not wearing a sundress; therefore, it must be raining cats and dogs. What kind of logic is this? Maybe the sundress belongs to my wife, and she isn't sharing. Such fallacies, as any child who takes his evolution with a pillar of salt knows, are examples of *affirming the consequent*. Lisle is hardly the first to be aware of the logical fallacy inherent in much of the experimental method. Historian of science Steven Goldberg similarly worries:

> There is a serious intellectual problem that Bacon and Descartes were well aware of and in fact so will Galileo, Newton, Leibniz and everybody else be aware of in attempting to argue that a scientific theory is true because it makes correct experimental predictions.[296]

Goldman agrees on the name of the fallacy. The fact of rain "proven" by either the wet sidewalk or my not wearing a sundress is an obvious example of affirming the consequent. About the fallacious approach, Goldberg lectures:

> That means that built into modern science is a flaw, so to speak. Built into modern science is a kind of a fracture zone that precludes certainty. We can never be certain because a theory works that it is true. We will see ... that many theories that were thought to be true because they worked were subsequently declared to be false.[297]

The Cats-and-Dogs Fallacy in Evolutionary Theory

That said, let's continue with our cats-and-dogs fallacies, this time applying them to evolution: If it were evolutioning cats and dogs, then cats and dogs would share a percentage of their

DNA; cats and dogs share a percentage of their DNA; ergo, it must be evolutioning cats and dogs. The fallacy is, of course, no different from my wet-sidewalk fallacies. Maybe God, or consciousness, or the inherent purpose of the universe, or an alien race writing the DNA and seeding the earth with it in the first place, failed to change handwriting between penning the chapters on cats and dogs and us. Maybe the quantum foam burped. Another cats-and-dogs fallacy: If it were evolutioning cats and dogs, then we would find no fossil dogs and cats in the Precambrian; we find no fossil dogs and cats in the Precambrian; ergo, it is evolutioning.

As the philosopher Karl Popper—who popularized falsification in the first place—knew, dogs and cats in the Precambrian would falsify evolution. Their absence proves nothing. Even real science— which has transformed the world with know-how and is therefore true if pragmatism and sophistry are true—has no logical ability to prove know-why with logical certainty, only to disprove it.

So what happens when we add two fallacies together? If the sidewalk is wet and I have no sundress, then is it twice as likely that it's raining cats and dogs? Do two fallacies make a truth? Is the sum of two fallacies two times as logical or two times as fallacious?

Consilience—Multiplying Fallacy

This brings us to the crown jewel of biomythology: consilience. The term was coined by Master of Trinity College, Cambridge, William Whewell (1794-1866). Whewell spliced together the Latin *com* for "with" or "together" and the Latin *silient* for "jumping." Thus, the "jumping together" or consilience of inductions about which Whewell wrote:

> The evidence in favour of our induction is of a much higher and more forcible character when it enables us to explain and determine [i.e., predict] cases of a kind different from those which were contemplated in the formation of our hypothesis. The instances in which this have occurred, impress us with a conviction that the truth of our hypothesis is certain.[298]

Certain? Paleontology is, for instance, "of a kind different" than, say, genetics. Thus if a hypothesis was consistent with both genetics and paleontology it would be, for Whewell, much worthier of a conviction of certain truth.

How does this work? Is consilience the jumping together of "inductions" or the addition of fallacies? Again, dogs and cats share DNA, but never in the Precambrian. Does this make evolution twice as true or twice as fallacious?

Maybe the problem is in the math. If sums demand addition, and products demand multiplication, and consilience is a product of the mind, and equivocation is close enough for evolution, then perhaps instead of adding the fallacies together to produce a sum, we should multiply them. Instead of our "truths" growing arithmetically like sustenance, they would grow geometrically like populations. As it turns out, this change from an addition to a multiplication sign is a veritable godsend for validating biomythology. Darwinian evolution multiplies the fallacies of genetics, biochemistry, paleontology, horseracing, anthropology, biochemistry, evolutionary psychology, and pig farming together until we find the true definition of consilience: multiplication of fallacy to a higher power.

Sophistry and the Universal Solvent

So, the next time a biomythologist brags about his or her reason and sound logic, the good skeptic can enjoy a good laugh. As for Bacon's alternative to syllogistic logic, induction, we will cover it in detail in our next chapter. Basically, induction is the faith that past experience allows us to predict the future—maybe. The truth of induction is that it predicts except for when it does not; the truth of fallacy is that it persuades except for when it does not. These truisms even the skeptic need not doubt.

12

The Road to Truth

Science becomes dangerous only when it imagines
that it has reached its goal.
—George Bernard Shaw

Truth

What is truth? Would we know it if it stood bound before
us? Is truth love? Could it be that simple? If not, what are the
alternatives? John Keats wrote, "Beauty is truth, truth beauty—
that is all / Ye know on earth, and all ye need to know." Will our
consciousness of truth, as surely as our consciousness of beauty,
fade into habit?

Is truth pragmatism? Was American philosopher and
psychologist William James (1842-1910) correct when he suggested
that truth is what it is good for us to believe? Is rationality truth
because it works, or if something works is it rational? Is faith truth
so long as it elevates moods, extends lifespans, or encourages
charity? Was American neopragmatist Richard Rorty (1931-2007)
correct when writing that the term *true* is "merely an expression
of commendation"? Is the term understood across cultures "just as
equally flexible terms like 'here,' 'there,' 'good,' 'bad,' 'you,' and
'me'"?[299]

233

Is something true if it succeeds? Alexander conquered a world, so is conquest truth? How long must the success last? Was Alexander's truth less true when his generals had divvied up his empire before time had jerked the flesh on his honeyed bones? Is truth a perfect correspondence between action and story from all perspectives? Is truth the passion turning story into belief? Is truth consensus? If all of science believes it, is it truth, or just belief? If the whole world agrees with it, is it truth, or just agreement? Is truth but a story concocted to redraw the borders of skepticism? Can we not have our God and Darwin too? Does truth demand we perforce doubt one or the other?

Plato, imagining the futility of searching for Truth in this world, saved the capital T for a higher world of eternal "Forms." Absolutes inspire—Plato thought—but exceed our reach. Engineers worship the straight line, but no such straightness exists outside of metaphysics. To transfer straight lines from the geometry above to the world below would require a miracle, a supernatural act. A straight line, however inspiring, viewed with an accurate enough measuring device, ceases to be straight; those contrary quarks refuse to stand on line.

We worship our absolutes; they guide us, inspire us. Even so, the seemingly straight lines of skyscrapers—like the face of the Divine in a baby's smile—are but reminders of the absolute. We live, therefore, on faith: that shroud tailored to shield us from the burning awe of mystery and sting of chaos, a shroud whose unvarying fabric fills us with awe, ecstasy, and hope to one day know Truth. Luckily, this chapter will make no attempt to define Truth. It instead guides the skeptic to navigate biomythology's construction project: the road to truth.

The Road to Truth

Take Interstate 90 from Boston to Seattle. It's a long road, the longest in the United States. The road to truth is longer, but you won't find it on Google Maps or MapQuest. Who owns the road to truth? Those who are more interested in using science to advance our minds than using their minds to advance science are pretty much agreed.

Consider the principle dogma of biomythology, "Science is the best approach to the search for truth, whether about mind or matter." The dogma is iterated and reiterated. Sam Harris asserts, "In the broadest sense, *science* (from the Latin *scire*, 'to know') represents our best efforts to know what is true about our world."[300] Michael Shermer agrees, "Although there is no Archimedean point outside of ourselves from which we can view the Truth about Reality, science is the best tool ever devised for fashioning provisional truths about conditional realities."[301] Jerry Coyne, also a believer, is a bit more conflicted:

> Darwin's theory that all of life was the product of evolution, and that the evolutionary process was driven largely by natural selection, has been called the greatest idea that anyone ever had. But it is more

235

than just a good theory, or even a beautiful one. It also happens to be true.[302]

Twenty-two pages later, Coyne admits:

All scientific truth is provisional, subject to modification in light of new evidence ... As we'll see, it is possible that despite thousands of observations that support Darwinism, new data might show it to be wrong.[303]

Most honest of all is Christof Koch. In *Consciousness: Confessions of a Romantic Reductionist,* he confesses:

Despite the hectoring of ... [critics], science remains humanity's most reliable, cumulative, and objective method for comprehending reality. It is far from fail-safe; it is beset with plenty of erroneous conclusions, setbacks, frauds, power struggles among its practitioners, and other human foibles. But it is far better than any alternatives in its ability to understand, predict, and manipulate reality.[304]

Even the skeptic could easily admire Koch. He is more of a seeker than a salesman. He has dedicated his life to the understanding of consciousness. Koch and researchers like him may one day prepare us to treat Alzheimer's disease, Parkinson's disease, head trauma, and stroke; they may learn to piece Humpty Dumpty back together again. However, Koch also shares the leap of faith upon which biomythology rests, "Because science is so good at figuring out the world around us, it should also help us to explain the world within us."[305]

Or not. The analogy could be false. The skeptic could reasonably suspect that just as perceptions in the universe of air may provide illusions in the universe of water, so may ideas based on perceptions in the universe of air provide illusions in the universe of mind. Universals are great; the problems arise for the skeptic only when we mix our universes.

Harris's "best efforts," Coyne's "subject to modification," and Koch's "erroneous conclusions, setbacks, frauds, power struggles" only reinforce that the road to truth is not the same as its destination. As for Shermer's "provisional truths about conditional realities"—to the skeptic it sounds like Latin for "the check is in the mail." Dawkins nevertheless reassures us, "The fact that some widely held past beliefs have been conclusively proved erroneous doesn't mean we have to fear that future evidence will always show our present beliefs to be wrong."[306] Thus, Dawkins denies induction applies to induction. Inductive success but not inductive failure is part of an ongoing pattern. The past predicts future successes, but not future failures. Successes are universal; failures are not. Even a skeptic could get confused.

The Turning of Centuries

Should the skeptic share Dawkins' optimism that induction will defy induction and fail to fail again? Is induction like scientific theory: true except for when it is not? Compare scientific truth in 1900 to now. In a lecture given on Friday, April 27, 1900, the grand old man of physics, William Thompson (1824–1907), the same who helped popularize "absolute zero" (a mystical land in which motion stops and the quarks freeze over), saw only "two clouds" obscuring "the beauty and clearness" of his "dynamical theory" about the motion of light waves moving through the ether. Within five years, light waves and the ether had both disappeared thanks largely to a fellow named Einstein. In 1900, the universe was stable; now it's expanding. In 1900, the heavens contained millions of clouds of gas; now those clouds are galaxies. In 1900, the stable arrangement of continents was assured; now the continents are drifting. In 1900, time and space were absolute; now they are relative. Reviewing this list, we are not talking about fine-tuning the system. These were monumental changes ushering in a new world of universals for a new universe, a new century. The real question is, by 2100, will we be living in yet another new world and universe? Does it help that some are as passionately certain about things now as William Thompson was in 1900? As philosopher Thomas Nagel cautions us in *Mind and Cosmos: Why the Materialist Neo-Darwinian Conception of Nature is Almost Certainly False*:

> It may be frustrating to acknowledge, but we are simply at the point in history of human thought at which we find ourselves, and our successors will make discoveries and develop forms of understanding of which we have not dreamt. Humans are addicted to the hope for a final reckoning, but intellectual humility requires that we resist the temptation to assume that tools of the kind we now have are in

principle sufficient to understand the universe as a whole.[307]

Cook's Test of the Miraculous

With our minds on our present limitations, it is useful for the skeptic to view a maxim from the skeptic of skeptics, David Hume: "No testimony is sufficient to establish a miracle unless the testimony be of a kind, that its falsehood would be more miraculous than the fact which it endeavours to establish."[308]

Today biomythology has adopted the pretentions once reserved for religion. Playing catch-up, I have modified Hume's maxim to fit the miracles of science rather than the miracles of religion: *No evidence is sufficient to establish a scientific theory unless the theory be of a kind that its falsehood would be more miraculous than the absence of conflicting discoveries in the unending future.* Today over 90 percent of the genome is "junk," and 90 percent of the matter in the universe is "dark." If *junk* and *dark* are synonyms for our ignorance, then our ignorance is no less than ten times greater than our knowledge. What chance, therefore, has the truth of today's theories against the ravages of tomorrow's discoveries? Before we can with certainty say that known forces and processes prove a theory's truth, we must be able to say that no currently unknown forces or processes could prove its falsity—or we must take the theory's truth on faith. If truth is an absolute, then justification for true belief must be no less absolute, or truth has been reduced to faith.

Sophistry Versus Platonism

The good skeptic should consider that some arguments never die. In ancient Greece, the Sophists were itinerant educators, traveling from city-state to city-state, teaching rhetoric and that consensus was formed by persuasion. The Sophists were relativists: what was true in Athens was not necessarily true in Sparta. Plato fought against relativism, championing absolute truths. Professor Goldman, in his lecture "Knowledge and Truth are Age-Old Problems," explains:

> In Plato's dialogue called 'The Sophist,' Plato talks of the battle between the gods and the Earth giants. It's a battle. It's war over who is to define philosophy—who is to define reason, knowledge, truth, reality. On the side of the gods, so to speak—this is of course Plato and Socrates' side—on the side of the gods are those who hold that knowledge is absolutely distinct from belief or opinion—that there is a qualitative difference between knowledge on the one side and beliefs or opinions on the other. Knowledge refers to that about which one cannot be wrong. Beliefs or opinions refer to that about which one could be wrong. That's why they're called beliefs and opinions. Knowledge is not a matter of belief or opinion. Knowledge is what is true.[309]

The battle continues today, but the opponents have renamed themselves; now we have realists and instrumentalists.

240

Realism Versus Instrumentalism

It is one thing to know how to work a smartphone. It is quite another to know how a smartphone works. The first may hardly require a preschool education; the other, a degree in engineering. So the question arises: does real science allow us to know how to work the world, or does it allow us to know how the world works?

On this question, philosophers of science divide into two groups—the realists and the instrumentalists. The realists contend science tells us how the world works; they argue that because real science has inspired technology to revolutionize the world, its explanations must faithfully describe reality and be "true." To realists, their stories are not just artifacts of human minds sculpted from a common ancestry with apes. To the realist, there really are flesh-and-blood quarks, not merely a product of unnatural levels of speed and energy multiplied by the expectations of math and faith. Realists similarly promote that anthropomorphic Nature—personified, reified—selects. Naturally!

The instrumentalists propose that scientific theory is an instrument for controlling the world, not abolishing mystery. For the instrumentalist, the beauty of magical know-how, not the truth of mythical know-why, is enough. Science tells us how to work the world, not how the world ultimately works. Theories are ideal. They exist only in our minds, no matter how often they appear to be exemplified or instantiated in nature.

The instrumentalists are the new Sophists arguing against absolute truths in scientific explanation. Formulas work or they don't work, but explanations are opinion, consensus, not truth. Instrumentalism returned to fashion in Copernicus's book *On the Revolutions of the Celestial Spheres*. Copernicus wisely waited until he was upon his deathbed to publish, because a mattress is more comfortable than a stake and a comforter more comforting than a fire. In *Revolutions'* preface, Andreas Osiander, sharing Copernicus's compunctions against martyrdom, advocated that a sun-centered universe predicted eclipses and the tides, but that was it. The theory's only truth was in its predictions, not that the earth actually revolved around the sun. As science historian Frederick Gregory writes:

> Osiander's position about the motion of the Earth has been cited as an early example of a philosophical position known as anti-realism [a kissing cousin of instrumentalism]. In this view, scientific theories would be regarded as assumptions that are adopted because of the usefulness, not because they represent nature 'as it really is.'[310]

The instrumentalist's view of science is ably captured in the preface of Brian Cox and Jeff Forshaw's *Why Does E = mc²? (And Should We Care?)*:

> Equally important, we want every reader who finishes this little book to see how modern physicists think about nature and build theories that become profoundly useful and ultimately change our lives ... Einstein may be forced to give way to an even more

accurate picture of nature. In science, there are no universal truths, just views of the world that have yet to be shown to be false. All we can say for certain is that, for now, Einstein's theory works.[311]

This is why Cox and Forshaw did not name their book *Einstein's Theories Are True*. Instrumentalists are concerned with know-how, with the magic needed to improve the world; realists—especially those without a technology capable of working the world—prefer the shouts of "I believe!" Evolution by natural selection is true! The skeptic can imagine the shouts of "Praise Darwin."

Nevertheless, the skeptic may wonder why some insist on substituting words like *true* or *fact* for *probable* or *probability*. Why build arguments on equivocation? Most of us, when we hear the word *truth* connected to a grand theory, think of a grand Truth, a higher unchanging Truth, not a Truth that is capitalized merely because Darwin is capitalized—except in Georgia where, as we have seen,[312] stickers promoting instrumentalism are banned from high school biology texts. Socrates would be proud to know that there is still a separation between philosophy and state and that those who philosophize can still be separated from the state—drink your hemlock and be quiet, Socrates.

Perceptual Truth Versus Theoretical Truth

It is one thing to use the word *true* for perceptions like "bricks feel hard." It is entirely different to apply the word to theories. To suggest that the theories of physics are "real" because your hand hurts when you hit a brick wall is kind of like saying, "Put two sandwiches, three apples, and a natural selection in the picnic basket." Except perhaps when Berkeley's ideas are applied to the sounds of unheard falling trees, ideas and perceptions are not in the same category.

Dawkins disagrees, consistently elevating theoretical truth to the level of habitual perceptual truth:

> A scientific theorem has not been—cannot be—proved in the way a mathematical theorem is proved. But common sense treats it as a fact in the same sense as the 'theory' that the Earth is round and not flat is a fact, and the theory that green plants obtain energy from the sun is a fact. All are scientific theorums [*sic*]: supported by massive quantities of evidence, accepted by all informed observers, undisputed facts in the ordinary sense of the word.[313]

The globe has been circumnavigated by Magellan, orbited by John Glenn, and seen by all of us on TV. At this point, the approximate roundness of the Earth is less an idea than a perception. It is a fact that many a green plant has failed to thrive without sunlight and that *energy* has been a very fruitful scientific concept. Even so, as Richard Feynman confessed in *The Feynman Lectures on Physics*, "It is important to realize that in physics today, we have no knowledge what energy is. We do not have a picture that energy comes in little blobs of a definite amount."[314] What we call energy today may therefore be understood in quite a different way in the future.

The skeptic may correctly ask, "If the world and the universe are not safe from the construction projects on the road to truth, what is?" To proclaim that "evolution is true" is to proclaim the limits of scientific creativity and discovery. Can truth be reduced to probability? Can truth be reduced to consensus? For the skeptic's benefit, we will examine these questions next, one at a time.

Induction

Consider a syllogism that is generally implied by those touting the wonders of induction: *All pragmatism is truth; induction is pragmatic; therefore, induction is truth.* The argument is valid; therefore, it is true if its premises are true. But, are they? Is pragmatism truth? And, what exactly is meant by the word *induction?*

To quote Sir Francis Bacon, "The mind loves to leap to generalities so it can rest."[315] To justify this laziness, (allowing the substitution of Nature's bad habits for the agreements of deductive logic), Bacon adds, "But we reject proof by syllogism, because it operates in confusion and lets nature slip through our hands ... we regard induction as the form of demonstration which respects the senses, stays close to nature, fosters results and is almost involved in them itself."[316]

Recall another syllogism from our "Fallacy to a Higher Power" chapter, "Darwin is a man; all men are fools; ergo Darwin is a fool." Or, fallaciously, "If it were evolutioning cats and dogs, the fossil record would be ordered; the fossil record is ordered; ergo, it must be evolutioning." Bacon knew that a syllogistic science could only be fallacious, so he had to settle for induction.

Induction is the prayer that the past will, most probably, predict the future. We saw the sunrise yesterday; we will see it rise tomorrow. As we discussed in our first chapter, a coin's run of heads will not end; if ten flips in a row come up heads, so will the eleventh. If the World Trade Center towers over New York City for 10,000 days in a row, then it will tower so long as the laws of engineering apply. Sadly, no matter how strong our faith in probability, we cannot, with certainty, predict the future.

Breaking Our Hume Habit

As we also suggested in our first chapter, philosopher of the Scottish enlightenment David Hume (1711-1776) warned against getting too comfortable with the presumption of the uniformity of nature. In his book *An Enquiry Concerning Human Understanding,* Hume wrote, "If there be any suspicion that the course of nature may change, and that the past may be no rule for the future, all experience becomes useless, and can give rise to no inference or conclusion."[317] Hume maintained that cause and effect were beyond reason and belonged, instead, to habit. As we have seen with the floating finger experiment, we have every reason to suspect the course of nature may change from one world of perception to another. Habit, though close enough for cause-and-effect perception in a familiar environment, may fail brutally in new environments.

As we have said, there is no immutable reason to suppose that the lessons of science in the physical world may be the rule in the world of the mind. There is no reason to trust biomythology. Nevertheless, kicking the habits of cause and effect and the future mimicking the past is even harder than kicking the habit of smoking. In this essay, we will require neither. Rather than embrace Hume's heroically delightful skepticism and attempt to lower induction to habit, we will elevate induction to its true stature: faith.

Monsters Swallow Induction Whole

Faith in the uniformity of natural law has created the miracles in physics. But all inductions are not equal. Biology is not rocket science. Biomythology (by definition) is not *real science* at all.[318]

Organisms routinely defy induction, stubbornly disobeying the rules. If induction applied to biology as it applies to physics, we would not have to count the fingers of newborns. Abhorring the embarrassment created by supernumerary digits and other anomalies, biologists routinely disparage the exceptions to their inductive prejudices as "monsters." Quasimodo rang the chimes at Notre Dame, not the hearts of biologists.

A recent neuroscience paper reported a little girl who suffered a cerebral mudslide before her seventh week of embryonic gestation.[319] As a result, she had only half a brain. Induction predicts that when gazing straight ahead, human beings see the right half of the world with the left brain and the left half of the world with the right brain. The little girl, having no right brain, nevertheless could see portions of the left half of the world when looking straight ahead. Her disrespect for induction was inexcusable because it drove home the point that, in biology, induction means only that things are normal except for when they are not—nothing on which to bet your house. In inductive science, truth reduces to probability. The good skeptic should suspect that confusing probability with truth is like betting on a horserace without a finish line. Knowledge of who won requires certainty—the photo finish, not the odds. To the skeptic, at least, anything with less than a 1.0 probability crossing the finish line hardly qualifies as certain knowledge. Probability insures that gambling casinos and insurance companies will win— barring card counters and other natural disasters. Probabilities can never predict if you personally will win or lose any more than quantum physics predicts the individual players in the half-life. Applying group probability to individuals is logical for groups, not for individuals. Your 99 percent chance of dying from your terminal cancer is erased the moment you are hit by a truck. To conclude with one last gambling metaphor: probability is no more certain than arm wrestling with a one-armed bandit. Truth is no less certain than leaving with the jackpot.

Inducing Consciousness

It only gets worse when we come to biomythology and the *scientific* control of consciousness. Add consciousness to biology and be prepared to flush all predictions goodbye. For the person who is fully conscious, a constant stimulus need provide no consistency of response. A lifetime of training patients to control visual consciousness has convinced me that the visually adept can see the same drinking glass as blurred or clear, double or single, half-empty or half-full. Consciousness demolishes induction as surely as the consciousness of scientists and military strategists demolished Hiroshima. Outside of pharmaceuticals and brain trauma, the closest that science has come to controlling the mind is in addicting rats or gamblers to pull a lever or run the maze of a casino floor. However, even such stimulus-response performance is unreliable. Raise consciousness and volition may break even the bonds of addition. We can predict when Pavlov's dog will drool, but not when the animal will get smart enough to eat the fool ringing the bell. The probability that yesterday predicts today serves as *near truth* in physics, *half-truth* in biology, and *myth* in the control of minds.

Put another way, induction posits universals, and universals are often hard to come by in this universe. The speed of light in a vacuum may be a universal, but the universe contains darn few absolute vacuums. Similarly, as I've noted any number of times, Newton's laws of motion are more like local ordinances; they can be enforced in neither the very small nor the very fast world. The skeptic would do well to suppose that if even the laws of physics are merely local ordinances, then the laws of biology are more like rules of thumb, and the laws of the mind are more like madness—unenforceable suggestions waiting to be defied by the will.

Scientific Consensus Equals Political Science

The good skeptic recognizes that the creed of biomythology begins not with "We believe!" but with "We agree!" Real science defines truth by the strength of the building, the bridge, or the medication. Biomythology defines truth by the strength of the consensus. For instance, no one flicking a light switch stops to mention that "it is the consensus among the scientific community that electricity is true." So long as science sticks to the job of controlling nature, no one could care less about consensus. As we have written, when the word *scientific* is bolstered with the word *consensus*, your only certainty is that you are in the middle of a sales pitch. An appeal to consensus is tantamount to a confession of selling ideology, not science.

Natural selection is an excellent example. Since the principle part of nature it controls is grant money, natural selection must rely for its "truth" on appealing to consensus. Consider, for instance, the introductory paragraph of the Oxford *Evolution: A Very Short Introduction*: "The consensus among the scientific community is that" 1) the earth rotates around the sun, 2) there are billions of galaxies, 3) the universe is still expanding from its origin 14 billion years ago, 4) the earth is about 4.6 billion years old, and 5) "All present-day living organisms are the descendants of self-replicating molecules that were formed by purely chemical means, more than 3.5 billion years ago."[320]

That the paragraph is entirely rhetorical would be harder to doubt than the data it contains. The paragraph's authors, Brian and Deborah Charlesworth, both excel in the sale of—rather than the search for—truth. They are marketing evolution by "positioning" it with some of this year's accepted scientific agreements: *Evolution is as respectable, as convenient to use, as inexpensive, as easy to find, as any other current agreement in science.* How many of these agreements will see the light of the next century?

Darwin knew the danger of mistaking consensus for truth. He wrote:

> When it was first said that the sun stood still and world turned round, the common sense of mankind declared the doctrine false; but the old saying of *vox populi, vox Dei* [the voice of the people is the voice of God], as every philosopher knows, cannot be trusted in science.[321]

Stephen Jay Gould provided the above quote to introduce his own masterpiece of rhetoric, which at the time ran against the consensus of his day. Despite Darwin's and Gould's warnings, *consensus* about evolution is repeatedly invoked. *Evolution for Dummies*, relying, as it does, for "truth" on both statistics and consensus, assures us in the introduction, "99.9999 percent of scientists accept the theory of evolution."[322] How many of those figures are significant and how many are rhetorical, we'll touch upon in a future chapter.[323] As always, Dawkins's statement of the consensus is more colorful: "Evolution is a fact ... No reputable scientist disputes it."[324] None dare. To do so would be to admit disrepute.

"Today scientists have as much confidence in Darwinism as they do in ... microorganisms as the cause of infectious disease,"[325] Coyne assures us. Who can argue? Infection, by definition, is the infiltration of "disease-causing microorganisms." Coyne's trivial remark rivals "the survival of the survivors" for its tautological strength. No matter that our bodies, inside and out, house trillions of microorganisms—the same number as neuronal connection in the brain—and yet we are normally healthy. When microorganisms are not sufficient to cause disease, they are by definition not infectious; when resistance fails, the identical microorganisms earn the infectious title. Despite the tautology, would all scientists agree that microorganisms, not decreased or absent resistance, are the real causes of disease? Would none note that the same microorganisms Europeans held hands with decimated Native American populations? Was it the microorganisms or the missing resistance that caused the disease? Is Coyne offering rhetoric or science?

Sure, the skeptic should note that my rhetoric is no better, but I don't pawn off my invariably false analogies as science. I make no claims that my words are backed by *scientific* concepts. Science, when reduced to words and abducted to control culture and minds rather than the physical universe, is no longer real science, no longer

Bacon's quest "to recover the right over nature"; it is the science of persuasion, rhetoric, metaphor, propaganda, analogy, ideology. Why demean *real science* by confusing it with biomythology?

What Darwin Got Wrong

As we have discussed, Darwinism has been disputed by prominent evolutionists ever since Darwin proposed natural selection as the mechanism for the arrival of the fittest.[326] The sanest treatment of natural selection and consensus was recently offered in *What Darwin Got Wrong*. Authors Jerry Fodor and Massimo Piattelli-Palmarini, two avowed atheists who seem to be more interested in defending logic than materialism, write:

> We've been told by more than one of our colleagues that, even if Darwin was substantially wrong to claim that natural selection is the mechanism of evolution, nonetheless we shouldn't say so. Not, anyhow, in public. To do that is, however, inadvertently, to align oneself with the Forces of Darkness, whose goal it is to bring Science into disrepute. Well, we don't agree. We think the way to discomfort the Forces of Darkness is to follow the arguments wherever they may lead, spreading such light as one can in the course of doing so. What makes the Forces of Darkness dark is that they aren't willing to do that. What makes Science scientific is that it is.[327]

Darwinians are disturbed that proponents of creation science and intelligent design have taken Karl Popper's theories to heart and think it their creation-science job to falsify Darwin rather than prove that evolution is true. So long as consensus is perceived as truth and Pluto can be voted out of office, fighting the rhetoric of intelligent design with the rhetoric of Darwinism is politics, not science. When we hear the words *everyone thinks*, the skeptic is justified in assuming that no one is still thinking. Like Pavlov's dogs drooling, we are conditioned to laugh.

Laughing at Consensus

Artists have long exploited this laughter. Take the writings of Voltaire (1694-1778), Twain (1835-1910), and Shaw (1856-1950). In the first few chapters of *Candide,* Voltaire shows a Christian philosopher's "best of all possible worlds" replete with slavery, imprisonment, flogging, war, mayhem, excrement, pustules, pox, bankruptcy, tempests, shipwrecks, earthquakes, inquisitions, and perversions galore. In *Pudd'nhead Wilson,* Twain, having explained that slaves are too valuable to imprison, concludes, "As soon as the Governor understood the case, he pardoned Tom at once, and the creditors sold him down the river." In the play *Mrs. Warren's Profession,* Shaw shares what he perceives as the true credo of the capitalistic spirit, "I'll do wrong and nothing but wrong. And I'll prosper on it."

We laugh, not just because these authors were satirizing Christian charity, slavery, or capitalism. They were exploiting the disparity between consensus and truth—the consensus of humanity and the truth of their inhumanity. Natural selection somehow finds the eye of the public better than it explains the eye of the eagle.

We see consensus once again falling somehow short of imagined truth. We laugh. As Thomas Nagel warns us of the materialist neo-Darwinian conception of nature:

> I would be willing to bet that the present right-thinking consensus will come to seem laughable in a generation or two—though of course it may be replaced by a new consensus that is just as invalid. The human will to believe is inexhaustible.[328]

Tying together our discussions on the limits of induction and consensus, a Karl Popper quote serves nicely:

> Some who believe in inductive logic are anxious to point out, with Reichenbach [a philosopher (1891–1953) known for his dedication to empiricism], that 'the principle of induction is unreservedly accepted by the whole of science and that no man can seriously doubt this principle in everyday life either.' ... Yet supposing this were the case—for after all, 'the whole of science' might err—I should still contend that a principle of induction is superfluous, and that it must lead to logical inconsistencies.[329]

Conclusion

Michael Shermer tells us that "science is the best tool ever devised for fashioning conditional realities." Whether or not Shermer is correct, biomythology is vying to become the best tool ever devised for fashioning certain belief out of uncertain truth. In Karl Popper's words:

> Now it is far from obvious, from a logical point of view, that we are justified in inferring universal statements

from singular ones, no matter how numerous; for any conclusion drawn in this way may always turn out to be false: no matter how many instances of white swans we may have observed, this does not justify the conclusion that all swans are white.[330]

For years, "all swans are white" was a European philosophical cliché often confused with truth. Then, black swans were discovered in Australia. The skeptic's lesson from this example? If love[331] is truth, things are pretty simple: truth and the road to truth are one and the same. When speaking of biomythology, however, the skeptic should never confuse the road to truth with its destination. As for the little girl we mentioned who lost half her brain by an error in development, perhaps her loss will be Darwin's gain. Perhaps one day she will appear in the fossil record and her half-brain will fill the missing link between real scientists and biomythologists.[332]

13

Some Falsehoods Are More Equal than Others

Teapot Theory

Consider another tool for skepticism. In biomythology and the rhetoric of objectivity, all falsehoods are equal, but some are more equal than others. Consider a fable by philosopher Bertrand Russell (1872-1970):

> Many orthodox people speak as though it were the business of skeptics to disprove received dogmas rather than of dogmatists to prove them. This is, of course, a mistake. If I were to suggest that between the Earth and Mars there is a china teapot revolving about the sun in an elliptical orbit, nobody would be able to disprove my assertion provided I were careful to add that the teapot is too small to be revealed even by our most powerful telescopes. But if I were to go on to say that, since my assertion cannot be disproved, it is intolerable presumption on the part of human reason to doubt it, I should rightly be thought to be talking nonsense. If, however, the existence of such a teapot were affirmed in ancient books, taught as the sacred truth every Sunday and instilled into the minds of children at school, hesitation to believe in its existence would become a mark of eccentricity and entitle the doubter to the attentions of the psychiatrist in an enlightened age or the Inquisitor in an earlier time.[333]

Explicating the passage, Richard Dawkins assures us, "Russell's point is that the burden of proof rests with the believers, not the non-believers." And well it does. I am responsible for what I allow into my mind, and Dawkins for what he allows in his. Responsibility for constructing our private stories is the cost of free will. That I am somehow responsible for the burden of proving anything in Dawkins's private stories, however, or that I am duty-bound to import Dawkins' private stories into mine, is a fairytale on the order of Russell's teapot.

Bunny Theory

Suppose the teapot's diminutive size continues to defy the resolution of our telescopes, but we discover comet-size bunnies throughout the solar system, all orbiting elliptically. Based on the regularity of the observation, we advance bunny theory (BT), professing that all orbiting bunny comets travel in elliptical paths. Soon BT is described in high school biology textbooks and five days a week instilled into the minds of children as "highly probable." Hesitation to believe in BT would become a mark of eccentricity and entitle the doubter to the censor of peer review, if not the attentions of the Georgia court system.

What would happen, however, were we to find a bunny comet in a *circular* orbit behind the once-planet Pluto? Which would now be falser: teapot theory or bunny theory? Which is the greater falsehood, something that cannot be falsified or something that has been falsified? Only those who could produce mind-independent realities of the mind could possibly define *falsehood* as "that which cannot be falsified." Still, if, as Popper suggested, all scientific theory must be falsifiable, and if truth cannot be falsified,[334] then scientific theory cannot be truth. It is small wonder that biomythologists feel the compulsion to see some falsehoods as more equal than others.

Contradictions Are All Too Logical

Perhaps this explains a common rationalization employed by historians of science: Great theories are never really false. Newer theories just explain the world more fully. Newton's absolute time and space, the keynotes of classical physics, still allow us to predict the trajectory of missiles. The modern physics of Einstein and Bohr and Feynman has merely extended knowledge to inspire the transistor and, subsequently, the digital age. Time and space are simultaneously absolute and relative. Neither is really false—the story goes. Newton is just a mite less true. Who cares if the two theories contradict each other? Logic is overrated. As we have seen, logical deduction doesn't work for science, which is why Bacon had to sell induction, despite its illogic.

Local Laws of Science

Another interesting falsehood that is not a falsehood was bequeathed to us by Antoine Lavoisier (1743-1794), the father of modern chemistry. Lavoisier graduated with honors in the

259

commencement exercises of the Enlightenment. In 1789, five years before his execution in the names of progress and reason, Lavoisier drafted the Law of Conservation of Matter, which was voted on, passed, and remained the dogma of international science for over one hundred years. Lavoisier and his peers were understandably ignorant of nuclear reactions or what happens to matter as it approaches the speed of light. Genius is the ability to see past belief. Einstein saw past the beliefs of his brethren and declared that the ignorance of natural law provides every excuse for breaking it.

Today the Law of the Conservation of Mass has been reduced to a local ordinance, obeyed by everyday chemists working with everyday chemical reactions. The example shows us that scientific laws are like phone numbers without area codes. However well they work in their hometowns, they often become useless when dialed outside the region of the universe in which they were discovered. A law that works in chemistry but not in physics is evidently a half-truth, not a falsehood.

Worldview Trumps Logic

Newer science just explains the world more fully. Consider Ptolemy. If utility is truth, it is not really false that the sun revolves around the earth. Just as Newtonian physics works for day-to-day spaceships but runs astray for the very small or very fast, an earth-centered universe with a movable sun still explains perfectly well why the sun rises and sets and why your shadow may disappear at noon (unless you are on the sun). Only when you want to calibrate your global positioning system rather than your sundial do you begin to need Copernicus. Oddly, the same biomythologists who hope to save face by claiming that Newton's absolute time and space are close enough for spaceships are often the first to declare that Ptolemy produced the falsehood of falsehoods and that Copernicus was our savior. As I've said, in biomythology all falsehoods are equal, but some are more equal than others, especially if they provide ammunition against conflicting worldviews.

Superstition and Inspiration

In biomythology, *superstition* is the falsehood of falsehoods being less equal than any other is. The Ten Commandments and the Golden Rule—derived, as the materialist's story goes, from mysticism, not science—are always more false than discarded scientific theories. Former errors in science, it seems, inspired the current errors in science, which will inspire the future errors of science. However, can we elevate disproven scientific theory on the grounds that the falseness of former theories inspired the probable falseness we now enjoy? If inspiration is all, why not include God? He did the trick for Bacon and Descartes and Boyle and Kepler and Newton, if not Sam Harris and Richard Dawkins. Perhaps, like falsehoods, some inspirations are more equal than others.

David Cook

Fairy Tales

Are some fairy tales also more equal than others? As we have seen, empiricism is practice based on experiment and observation. Such observation is not without danger. Consider the tale of "Little Red Riding Hood" by the French author Charles Perrault, who also gave us "Puss in Boots," "Cinderella," and "Sleeping Beauty." Little Red, a country girl living in a village, was "the prettiest creature ever seen." She was carrying custard for her ill grandmother when she met a hungry wolf in the woods, who found out her destination. While Red was "gathering nuts, running after butterflies, and making nosegays of such little flowers as she met with," the wolf ran ahead, ate the grandmother, and crawled under the bedclothes to await dessert. Red arrived and, good empiricist that she was, began to examine the wolf who was wearing her grandma's nightclothes:

> "Grandmamma, what great ears you've got!"
> "The better to hear, my child."
> "Grandmamma, what great eyes you've got!"
> "The better to see, my child."
> "Grandmamma, what great teeth you've got!"
> "To eat thee up!"

Little Red's careful anatomical survey and the wolf's careful list of physiological correlates is a cautionary tale against obsessive empiricism. Excuse my fun, but is empiricism really enough to guard a story from falsehood? In the real world there are, after all, nuts and nosegays and grandmothers and custards and wolves and woods and villages and creatures of beauty galore. Is it fair to call a story containing so much truth a complete falsehood just because the wolf's language skills threatened to fall off the nether end of the bell curve?

Once upon a time in logic, as I misunderstand it, something was true or false, not true *and* false—not true and its contradiction also true. Consider two sentences: "Wolves who talk are monsters." "Charles Darwin was a transvestite." If a wolf talked, it would indeed be a genetic monster, so the first sentence is true. The second sentence, I would imagine, is false. Which rankles more, the overestimated language skills, or the unfounded characterization? Should we be more threatened by the possible transcendence of language or the certain mischaracterization (for better or worse) of local heroes?

In other words, which is the greater falsehood: a talking wolf, or Darwin in a tutu? Would the greater improbability of a wolf talking than Darwin cross-dressing on strolls at Downs make the talk falser than the walk?

As we discussed in "The Road to Truth," probability is grossly overrated. This celestial numbers game threatens to transform horseracing into mysticism because at the finish line all probabilities become *one*. So if all winners have a probability of one, how come the statistics fail to hold for losers? Is a loser at the finish line not as much of a loser if his odds were better before the race? Is a falsehood, once proven false, not equally false no matter what its former probability of truth?

Again, as with our bunnies and teacups, can we convict a belief because of lack of evidence and not a former belief that the evidence now proves wrong? Is it falser to violate a worldview than a view in the world? Where's the logic in that? If one day it is shown that the structuralists were right and natural selection/organismic capitalism does not explain the arrival of the fittest, as we have previously discussed, won't Darwin's and Perrault's tales be equal falsehoods despite their share of truths? Won't both be fairytales? Won't both Darwin and Perrault have to be taught in literature rather than biology? Darwin may have inspired those who followed, but what if Perrault inspired J. K. Rowling? Which is more likely to endure through the centuries, selfish genes or Harry Potter?

So why distinguish the stories of Charles Perrault from the stories of Charles Darwin? Does the metaphor of gradual, tiny, externally selected changes of form turning microorganisms into marching bands prepare us for life any better than the metaphor of a wolf in Grandmamma's clothing? Have the tales of Red Riding Hood, Cinderella, and Sleeping Beauty not inspired as many people as Darwin's stories? Sure, Darwin—providing that starvation isn't throwing the epigenetic protein switches of gene expression—may have gotten the finch beaks right, but Charles Perrault took great pains to point out how the wolf's ears, eyes, and teeth were fully adapted to their function in their ecological niche. Why raise one savvy empiricist above another? A falsehood remains a falsehood and a fairytale a fairytale no matter how many empirically *true* facts it contains—or so the skeptic might argue.

Imaginary Friends Defined

Consider another category of falsehood, the imaginary friend. Richard Dawkins claims that God is, at best, an "imaginary friend." He asks himself, "Is the imaginary-friend phenomenon a higher illusion, in a different category from ordinary childhood make-believe?"[335] He then answers himself (not always a good sign), "I suspect that the Binker phenomenon [Binker was Christopher Robin's imaginary friend] of childhood may be a good model for understanding theistic belief in adults."[336]

In other words, according to Dawkins, God—like an orbiting china teacup—is an imaginary friend. Fortunately, *we can define a word how we like for our own purposes, provided we do so clearly and unambiguously.* The skeptic may define the term *imaginary friend* as "a companion held in our minds on whom we depend, or have depended, and who doesn't (currently) exist in the physical universe."

Favorite Imaginary Friends

A transcendent God, falling outside the physical universe, fits both definitions: Dawkins's and mine. My definition, as the skeptic will see, is more devastatingly fluid. A number of theories have frequented the minds of scientists but no longer exist in the physical world. Let us review. The earth at the center of the solar system was once the imaginary friend of Aristotle and Ptolemy. Parallel lines were the imaginary friends of Euclid. The sun smack dab in the middle of the entire universe was the imaginary friend of Copernicus. Absolute time and space were the imaginary friends of Newton. Light waves were the imaginary friend of Huygens. The ether was the imaginary friend of Maxwell. The conservation of matter was the imaginary friend of Lavoisier.

Darwin's imaginary friend was organismic capitalism stacking the individual longer vertebrae of the giraffe's neck. Many still imagine they see Darwin's friend. Many don't. Many never did. As we've discussed elsewhere, as long ago as 1884, William Bateson, the inventor of the term *genetics* and founder of the *Journal of Genetics*, wrote of the preposterous stories surrounding Darwin's imaginary friend, "for on this class of speculation the only limitations are those of the ingenuity of the author."[337]

Natural selection was the imaginary friend shared by both Charles Darwin and Richard Dawkins. Dawkins dismisses all limitations in seeing this friend as "a failure of imagination." Imagination is indeed the key in natural selection, and no one has a better imagination than Dawkins, whose brand of natural selection remains an inspiring and brain-warming tale whose gaps can easily be filled with imaginative reasoning:

> The flagellum is a thread-like propeller, with which
> the bacterium burrows its way through the water. I

say 'burrows' rather than 'swims' because, on the bacterial scale of existence, a liquid such as water would not feel as a liquid feels to us. It would feel more like treacle, or jelly, or even sand."[338]

How's that for an imaginary friend, a bacterium with a knowledge and consciousness of how treacle (syrup) feels? The selfish personalities of the genes in Dawkin's *The Selfish Gene*, the finer sensibilities of treacle-feeling bacterium, Darwin's natural selection "daily and hourly scrutinizing," his wee beasties "charming the ladies"—all these, Dawkins would probably insist, are merely Darwin's *rhetorical devices*. That Dawkins and Darwin do not let their science get in the way of their rhetoric is the very point. I thank both for inspiring me.

The Relativity of Wrong

The most obvious rebuttal to this chapter would be to quote, out of context, from a passage by biochemistry professor and science fiction writer Isaac Asimov:

When people thought the earth was flat, they were wrong. When people thought the earth was spherical, they were wrong. But if you think that thinking the earth is spherical is just as wrong as thinking the earth is flat, then your view is wronger than both of them put together.[339]

Asimov was talking about action in the real world, not universals, not reality in any absolute sense. For Asimov, if you are walking on a one-mile plain, a flat earth theory is good enough. If you are measuring the various diameters of the earth with accuracy, a spherical earth is wrong.

Close Enough for Horseshoes

Another way the skeptic may look at Asimov's passage is that science is like playing horseshoes. Close may be good enough; closer may be even better. Thus, some falsehoods are indeed more equal than others. The world of absolutes, of universals, is more like baseball. A foul ball is never fair; all foul balls are equal and none is more equal. Choose your analogy, but stick with it.

Real science is like horseshoes; some falsehoods are more equal than others. Biomythology, speaking of *truth*, confuses the rules of horseshoes with those of baseball, one minute ranking falsehoods, the next minute espousing, "Natural selection is a home run." "Foul and fair" is reserved for *Macbeth*. It is small wonder that biomythology, with its mixed metaphors and false analogies, has about as much chance to tell us what it means to be human as a foul ball has to become a home run.

In the world of absolutes, in the ideal world of the private stories we call *mind*, in which eternal Forms and equal signs are no longer approximations, foul is foul and fair is fair and all falsehoods are equal. As we pass from the world of mind, in which walls are mostly open spaces, into the world of commonsense perception and action, in which walls are hard, the measure of our success is our control of the world, not our rhetoric. If certain truth depends on certain proof, there will be no truth until we come to the end of time and the certainties congeal to cease changing. If falsehood is falser when it *cannot be* falsified than when it *has been* falsified, then there is no need to waste our remaining time on logic. Close is good enough for horseshoes, but not eternity. The good skeptic recognizes the difference between homeruns and horseshoes and doubts anyone who confuses the two.

14

The Madness in Shakespeare's Method

Excuse me now as we break for a little comic relief with which to sharpen our skepticism as we witness the universal solvent of rhetorical reason at work. As we continue our survey of the rhetoric of objectivity, another question arises. Which is mightier in evolving truth about what it means to be human: art, religion, or biomythology? The answer may surprise you.

In England at least, Sir Francis Bacon is credited with having invented the scientific method. Some even credit Bacon with having written the plays of William Shakespeare. (Insert laughter here.) What passes as scientific experimentation was no surprise to anyone when Bacon was still little more than your average courtier trying unsuccessfully to win his way into Queen Elizabeth's favor. So who really discovered the scientific method?

Method Acting

Around 1599 or 1600, when Modernity was dawning and Shakespeare wrote *As You Like It*, he already knew that studying the human mind was precarious: "All the world's a stage, and all the men and women merely players." Thus, unless we are all method actors generating our performances from our emotions rather than the practiced art of portraying characters, how do we, with certainty, know anything about the minds of others?

Compare how you act in front of your best friend, your children, your mother-in-law, and your boss. Which performance portrays the real you? What is the solution to the study of what it means to be human? "There are more things in heaven and earth, Horatio, / Than are dreamt of by your philosophy." With philosophy wearing thin, Shakespeare invented something new: the scientific method.

The Method in Hamlet's Madness

As *Hamlet, Prince of Denmark*—written or at least revised around 1600, two decades before *The New Organon* ("organon": Aristotle's tools of logic) was a twinkle in Francis Bacon's eye— begins, Hamlet's Uncle Claudius has murdered Hamlet's father and married Hamlet's mother. Hamlet is dismayed by his mother's hasty remarriage but does not yet suspect the murder, for Claudius's mind is to be found in neither his words nor his behavior.

Disturbed by the yet-to-be-codified science of behaviorism— the assumption that behavior, like the label on a can of beef stew, suggests the contents of the can, or mind—the ghost of Hamlet's father appears. Just to keep things objective, Shakespeare carefully confirms the ghost's existence by the observations of independent sentries and Hamlet's university-trained friend, Horatio, not to mention repeated sightings by the audience. True, later in the play, Hamlet's mother is too embarrassed to see the ghost, but then she possibly—being distracted by the King's brother—had looked right through him when he was alive. In any case, the ghost acquaints Hamlet with the facts of his murder by poison by good Uncle Claudius:

> Sleeping within my orchard,
> My custom always of the afternoon,
> Upon my secure hour thy uncle stole
> With juice of cursed hebona in a vial,
> And in the porches of my ears did pour
> The leperous distillment.

What to do now? Observation by multiple observers is obviously not enough. They could all be the subjects of some group hallucination. Maybe the ghost is an illusionist, like David Copperfield. If Hamlet's flesh-and-blood uncle may "smile and smile, and be a villain," can Hamlet have any greater certainty in trusting the words and countenance of a ghost? What if the ghost were the devil in disguise? Justifiably skeptical of the real intentions of both the living and the dead, Hamlet creates the scientific method and applies it to consciousness to become the first biomythologist.

Hamlet painstakingly designs his *Mousetrap* experiment, arranging for his old acting chums to perform a play at his uncle's court. Hamlet adapts the play so that his father's murder is reenacted in pantomime and word. During the performance, Hamlet watches his uncle's face, and for the purposes of verification, has his friend Horatio make an independent assessment. Oh, sweet objectivity!

The stage murder proceeds; the uncle flinches; the truth is revealed; Horatio independently confirms it! Hamlet's experiment catches the consciousness of a king, and all hell breaks out.

The results of applying the scientific method to consciousness are, as always, tragic. The experiment complete, Hamlet abandons his skepticism, and all too soon his fickle mother, Gertrude; his treacherous uncle, Claudius; his fatuous college buddies, Rosencrantz and Guildenstern; his best girl Ophelia; Ophelia's sententious father, Polonius; Ophelia's primrose-path-prancing brother, Laertes; and Hamlet himself all die violent deaths. Such is the efficacy, Shakespeare discovers, of abandoning skepticism for scientific knowledge about the human mind.

By the time Shakespeare writes *Othello* and *Macbeth*, he has disabused himself of any possibility of the scientific method capturing consciousness. He offers Othello no experiment to see into Iago's mind as the villain plots the death of Othello's "world of kisses." Shakespeare offers King Duncan no experiment to study Macbeth, the smiling host who will soon murder the good king in his sleep. Shakespeare is resolved to the impenetrability and boundless potential for surprise from other minds—scientific method, behaviorism, or not.[340]

Rationalizing the Scientific Method

Whether or not Shakespeare is responsible for the invention of the scientific method, he is responsible for another invention, at least according to scholar Harold Blum in his *Shakespeare: The Invention of the Human*. But, what experiment proves it is safe to mix the two inventions? What study shows that Bacon's objectivity outweighs Shakespeare's subjectivity as Macbeth's soliloquy describes daggers floating in the air? Bacon himself was aware that an "art cannot be condemned when it is itself both the advocate and the judge."[341] Thus, just as philosophy could not be used to condemn philosophy, neither can science be used to condemn science. Or can it?

Anecdote Is Not Science

Did you hear the one about the traveling evolutionary psychologist who slept in the farmer's barn? Such a story—while undoubtedly unfit for the ears of children—is known as an anecdote—a "narrative of an amusing or striking incident (originally an item of gossip)." Anecdotes are good enough to amuse guests at parties, but scarcely good enough for science.

As covered in our introduction, I'm an optometrist offering vision therapy—a kind of physical therapy for the eyes. If there is one thing I have learned, it is that when it comes to science, it doesn't matter how many anecdotes you can tell, how many success stories you offer, how many lives you change; without a well-controlled study, anecdotes amuse but prove nothing.

Defining the Scientific Method

Richard Dawkins loves to tell stories to children:

> Indeed, to claim a supernatural explanation of something is not to explain it at all, and even worse to rule out any possibility of it ever being explained. Why do I say that? Because anything 'supernatural' must by definition be beyond the reach of a natural explanation. It must be beyond the reach of science and the well-established, tried and tested scientific method that has been responsible for the huge advances in knowledge we have enjoyed over the last 400 years or so.[342]

So what is this "tried and tested" method responsible for all these advances? Unfortunately, scientists have yet to define conclusively the scientific method. Repeatability is obviously no criterion, for no one can repeat the experiment of turning matter into life without a jumpstart from DNA. Observation is not a criterion, for no one has observed one family of organisms turning into another family of organisms. Paul Feyerabend does, perhaps, the best job of pinpointing exactly what science is. In *Against Method*, he uses his encyclopedic knowledge of scientific history to conclude about the advance of science that "the only principle that does not inhibit progress is anything goes." As Feyerabend elaborates:

> The idea of a method that contains firm, unchanging, and absolutely binding principles for conducting the business of science meets considerable difficulty when confronted with the results of historical research. We find, then, that there is not a single rule, however plausible, and however firmly grounded ... that is not violated at some time or other.[343]

Although we can't agree on what exactly the scientific method is, we can agree that it works: Richard Dawkins tells us so. Let's examine a few prospective, controlled studies proving the point.

Proving the Scientific Method

Imagine this. There are no prospective, multicenter, controlled studies proving that the scientific method works—only anecdotes. "Nonsense!" you say. "What about Copernicus, Galileo, Kepler, Newton, Maxwell, Einstein? Don't scientific studies support gravity, a sun-centered cosmos, elliptical orbits, electromagnetism, and quantum theory?"

Of course they do. These are wonderful anecdotes all, but they leave questions. Where is the study showing us that the scientific method and not something else produced the discoveries? Is comparing the success of the scientific method in real science to the success of the scientific method in biomythology like comparing the success of weight training for football to the success of weight training for billiards?

True, humans are increasing their control of the physical universe, but biomythologists aren't interested in controlling anything but our minds. Rather than promote how their theories have made us better parents, lovers, saints—whatever—they promote the wonders of "the scientific method." So where are the studies supporting the scientific method itself, especially when applied to consciousness? To cite the successes of Copernicus, Galileo, Kepler, Newton, Maxwell, and Einstein is to offer nothing more than anecdotal evidence—evidence about the physical world, at that, not the mind. As biomythologists agree, when it comes to the human mind, it doesn't matter how many anecdotes we offer or how many success stories we provide; nothing changes. Anecdotes of scientific successes do not prove the method itself any more than anecdotes about the successes of the latest panacea prove the efficacy of the panacea.

Are We Ignoring the Failures?

Think of how many times the scientific method has produced falsehoods rather successes. A 300-page book of them is provided by John Grant in his *Discarded Science: Ideas that Seemed Good at the Time.* What is the possibility that we are paying attention to only the successes and ignoring the failures? What is the possibility that the successes could have occurred by chance or by using some discipline other than science? Think of how education and communication allow our creations to feed off of each other. Is it the scientific method or improved education and communication making the changes?

Intimacy

Another factor to be considered is intimacy. Just as intimacy, honesty, and the sharing of confessions with the group helped spearhead Christianity, so intimacy, honesty, and the sharing of exact procedures and discoveries could have spearheaded science. One of the genuine strengths of Bacon's ideas on experiment relies on his demand for telling all:

> We give a frank account of the method of the experiment we used; so that after we have revealed every move made, men may see any hidden error attached to it, and may be prompted to find more reliable, more meticulous proofs (if any exist).[344]

Experimental intimacy unfortunately allows for an occasional illicit affair. If the performed experiment doesn't perform as prejudice dictates, it need not be published; there is no necessity to miss and tell. The various artifacts of the experiment can be rearranged until they align with expected truth. Such manipulation won't launch rocket ships or light our homes, but the approach is close enough for persuasion, promotion, and biomythology.

Control Groups

Is it mere intimacy or the scientific method resolving problems? How do we know the hype surrounding the method doesn't provide the scientists with expectant faith, that it is not the method at all but the hope the method engenders that inspires the hard work and phenomenal cooperation that leads to success?

In biomythology, we explore the efficacy of a procedure by comparing the results of one group—"the experimental group" using a particular protocol—to the success of another group—"the control group" not using the protocol being tested. Even though there is no agreed-upon protocol for exactly how to do science—some stressing observation, some stressing experiment, some stressing objectivity, some stressing subjectivity, some despairing at ever demarking exactly where science ends and pseudoscience begins—what is any self-respecting study without a control group? Were we to test the efficacy of the scientific method in revealing, not only how matter interacts, but what it feels like to be human, what it means to be human, what group could serve as a control?

Religion Versus Science

Biomythologists assure us that, at best, religion is inert. What, therefore, could be a more suitable control group for our experiment? Andy Stanley, for instance, is the pastor of a local metro-Atlanta megachurch. Weekly he implores his parishioners to love one another (and us) as Jesus loved them. Suppose we assigned Andy's 36,000 parishioners and their children to the control group and assigned the 36,000 members of the American Psychiatric Association and their children to the experimental group. On a scale of happiness, hope, and certainty, which group would score highest, the science-trained psychiatrists or Andy's love-God-neighbor-enemy-trained Jesus followers? When it comes to the mind, when it comes to producing happiness and hope, which is better: religion or the application of the scientific method to consciousness? If biomythology cannot outperform "worthless" religion, is it less than worthless? Where are the studies to decide?

The Universal Solvent

As always, the universal solvent of rhetorical reason can be used to wash away the problem. Here are a few suggestions: 1) "Anything goes" is hard to falsify. The scientific method may not qualify as science, but even so, we can redefine science to include religion when necessary and exclude religion when necessary. 2) We can argue that the scientific method, like God in the infinite regress, is the prime proof unproved, that the scientific method is just another mysterious incomprehensible brute fact like charge and mass and space-time. 3) We can revert to aphorism, claiming that "science was created following the rules of science no more than chess was created following the rules of chess." 4) We can throw up our hands and admit there is no experiment proving experiment, no method proving the scientific method, no controlled study proving that science is superior to religion (or Shakespeare) in apprehending or inspiring the minds that make us human.

Creativity Versus Objectivity

Where may the skeptic assume that art ends and science begins? Bacon tells us to observe, observe, observe, and then experiment. Shakespeare observed, but more importantly, he created. What do we admire more: the subjectivity that created

Newton's gravity and Einstein's relativity, or the objectivity that validated them? Who *discovered* it? Who *invented* it? Who *created* it? Who *imagined* it? Why so much fuss about priority? Are we in love with the objectivity coordinating scientific consensus, or the transcendent subjectivity of some of the most gifted individuals in history riding on each other's shoulders to create, invent, and imagine the modern world?

Is Dawkins really awed by the scientific method or by Darwin's creativity and rhetorical skills? Is the strength of the scientific method in its objectivity or subjectivity? Which is more important—the confirmation process, or having something so good that it scarcely needs to be confirmed? Did the public need science to confirm the efficacy of the light bulb, or is such objectivity needed only to establish pet theories such as Darwin's? In truth, as the pages of the National Academy of Sciences in the 1880s confirm, Edison's light bulb was voted in by consumers long before the theories about the light bulb were voted in by scientists:

> Becoming restive at the slow progress of discovery, the inventor has himself assumed the *rôle* of investigator; and the results of his researches appear in the records of the patent office ... In consequence, the discoveries upon which many of the most important scientific investigations of the day rest, will be searched for in vain in scientific literature. The telegraph, the telephone, the electric light are inventions which illustrate the fact now stated, in an eminent degree.[345]

David Cook

So, outside of the endless anecdotes, what if the researching skeptic can find no study confirming the value of the scientific method in shattering the mind and studying its shards, no study confirming that the objectivity of scientists is better at expanding consciousness than is the subjectivity of artists or spiritual seekers? Does this leave us sitting in mystery or point the way toward fideism, the doctrine that knowledge depends on faith?[346] Never fear. Mystery, transcendence, infinity, eternity, creativity, love—all ideally are in our minds. All exist in spiritual space. All depend on subjectivity. How objectivity measures subjectivity, how the scientific method— proven or unproven—will solve the mystery of the mind, of what it means to be human, remains a mystery, an excuse for skeptics to throw a party. Still, we must have faith that reason will eventually succeed. Or not.

15

Rhetorical Numbers and Apes

Philosophy is written in this grand book—I mean the universe—which stands continually open to our gaze, but it cannot be understood unless one first learns to comprehend the language in which it is written. It is written in the language of mathematics.

— Galileo Galilei

Numbers are *scientific*—so *scientific!* In real science, numbers predict; in biomythology, they persuade. If the numbers run out, then there is always the magic of infinity. As our primer for skepticism continues, let's consider the use of numbers and the infinite in the rhetoric of objectivity.

The Father of the Scientific Revolution?

Some say that the universe is twenty billion light-years wide, meaning that it would take light—a particle or wave or ray, or whatever metaphor we chose—twenty billion years to wade across the universe. So if I were to suggest moving the center of the universe by eight light-minutes (the distance light travels in eight minutes rather than twenty billion years), then would you be impressed? Would you call me a genius? Would you suggest that for my proposal I had earned the right to be called the "The Father of the Scientific Revolution"?

That, of course, was Copernicus's achievement. He moved the center of the universe by eight light-minutes. Biomythologists, eager to show that Copernicus was way too smart to fall for the biblical story of Joshua, never mention that Copernicus had erred by a factor of, probably, billions. They never mention that the great scientist placed our sun not only at the center of the planets in our solar system, but also at the center of the universe! Never has anyone been made more right by being more wrong. Never has a group been more successful at selling uncertain truth.

The good skeptic may protest. Copernicus's achievement was in resurrecting an 1800-year-old tale told by Greek astronomer and mathematician Aristarchus of Samos (c. 310-230 BC) and giving it mathematical feet—no matter that the math, being based on Aristotle's perfect circular orbits, didn't work. Or, maybe you are protesting that Copernicus's achievement was saying that the earth revolved around the sun and that our sun and not our planet is the center of all things! Maybe you are impressed because once religion and Aristotle were at the top of the circle, and Copernicus gave a shove, and the circle began revolving to place science at the top and Aristotle and religion at the bottom—a revolution. Or, maybe, like any good skeptic you are impressed because Copernicus perfectly fits Nobel Laureate Richard Feynman's maxim: "Science is the belief in the ignorance of experts."[347]

Rhetorical Numbers

Whatever you are thinking, this is the point: all the numbers I provided—the eight light-minutes and twenty billion light-years and 1,800-year-old theory and factor of billions—were examples of *rhetorical numbers*: numbers used to persuade rather than predict. In this case I was purposely using numbers, which were correct enough in themselves, to malign a great scientist. As English statesman Benjamin Disraeli famously remarked, "There are three kinds of lies: lies, damned lies, and statistics."[348] And, as Scottish author Andrew Lang less famously wrote, "He uses statistics as a drunken man uses lampposts—for support rather than illumination."[349] In the words of my own aphorism on statistics, "The art of statistics rests in painting persuasion by the numbers."

The Statistics of the Human Heart

Numbers illuminate the facts of the world. Rhetorical numbers highlight rhetoric. Consider these facts: the adult male human heart weighs ten to twelve ounces; the adult female heart weighs eight to ten ounces. Now consider how the same numbers can be used to persuade: the adult male heart outweighs the adult female heart by an average of fifteen to twenty percent. Statistically speaking, men have bigger hearts than women do!

As we can see, the matter of the heart reduces to statistics; matters of the heart descend into rhetorical numbers. Rhetorical numbers are used to manipulate minds rather than matter.

Infinite Magic

Darwin's theory depended on the immensity of time. According to the story, natural selection takes tiny steps. Infinite time makes room for an infinity of steps, each carrying on where the previous step left off. The skeptic worries: if the steps are too tiny to see, can an infinite number of invisible steps replace an invisible God? If natural selection were to stack the giraffe's longer vertebrae, it would take time—lots of it! And, what if the giraffe had problems making the leap from two- to seven-inch vertebrae? No problem—a magical infinity of steps was available in between, flitting around like fairies, to create the seeming miracle.

A magical infinity of steps is the answer to every seeming irreducible complexity. What good is half a wing? Have faith! There was a magical infinity of steps in between the half and whole wing, each with a magical infinite advantage of its own—three-fourths of a wing made an excellent toilet plunger, the improvement in plumbing providing an evolutionary advantage. Falsify that. Whether or not we locate the toilet plungers in the imperfect geological record make no difference to our tale.

The Magic of Deep Time

Darwin understood that only the omnipotence of deep time could have allowed natural selection enough tiny steps to perform the miracle of constructing the ameba's butt and the eagle's eye from a common ancestor. It is little wonder that when the grandest of the grand old boys of physics, Lord Kelvin (1824-1907), calculated the cooling of the earth and declared (wrongly—radiation and the heat it generates having not yet been discovered) that the earth was not as old as Darwin imagined, Darwin referred to him as an "odious spectre."[350] *Why?* Kelvin was using science to kill Darwin's god of deep time as surely as Darwin was using science to kill Lord Kelvin's Lord.

Magical Infinite Thinking

Deep time (and the infinity of invisible steps it contains) is only one example of the new *magical infinite thinking*. The *deep pockets* of genes are another. As we have covered, William Bateson was one of the fathers of modern genetics. Bateson discredited natural selection. It is, therefore, only fair that Darwinian apologist Daniel Dennett call into question Bateson's perceptions:

> [Bateson] ... just could not imagine DNA. The idea that there might be three billion base-pairs in a double helix inside every human cell was simply not within the scope of his imagination. Fortunately, other biologists didn't share Bateson's pessimism, and they went on to discover just how the apparently miraculous feat of transmitting genetic information from generation to generation is achieved by some pretty fantastic molecules.[351]

Bateson, it appears, could just not conceive how deep the pockets of our genes are. Just imagine the billion base-pairs inside every human cell! Multiply that by 10^{14} or so cells per human and multiply that by your zip code and the number is very large, so large it can easily account for how genes are practically synonymous with life. Such is the magic of magical infinite thinking.

Magical Deep Tissue

Another example is the magic of *deep tissue*. The brain is complex. This year it has about 100 trillion connections. Multiply that deep tissue by seven billion people, and there's an infinity of magical thinking going on. If we multiply fourteen billion years of deep time by three billion base-pairs per cell by 10^{14} cells by 100 trillion connections in the brain by seven billion brains by your zip code, there ain't nothing we can't accomplish. Magical infinite thinking has replaced the need for God.

An Infinite Problem

There is a problem, the skeptic may notice, with all those brain connections and base-pairs floating around: with all that deep time, deeply hidden steps, deep pockets of genes, and deep tissue of brains (not to mention the deep zip codes of deep space), what are the odds that in the magical infinity of the future a new discovery could displace any theory that biomythology is currently selling? The skeptic knows enough to doubt when someone throws around magical infinite thinking for the purposes of rhetoric.

Is the Light Christian or Buddhist?

When the numbers run out in the dot-to-dot pictures of the cosmos, we have reached infinity. When it comes to infinity, many have claimed to see the light. The claims vary, however, on the nature of the light seen. Buddhists and Christians, for instance, share different perceptions. The keystone of Buddhism is that "we all suffer." The keystone of Christianity is that "we all sin." In the world today, there are five hundred million Buddhists. There are two billion Christians. If consensus is truth, then the vote is in—it is more fun to sin than to suffer.

The apologists for biomythology nevertheless love to point out that since Buddhism and Christianity have different statements on the light, both statements cannot be true and—biomythology ably extrapolates—since we cannot prove which viewpoint is true, we might as well regard both as false and throw out all other religions as well for good measure.

Is seeing the light any less problematic for scientists than the faithful? In the seventeenth century, Christiaan Huygens saw the light as waves; Newton saw the light as particles, or corpuscles, as he called them. In 1801, Francis Young performed experiments proving once and for all that light was composed of waves. Throughout the 1800s, when scientists saw the light, it was therefore waves. Then in 1905, Einstein, inspired by Max Planck, saw the light as composed of particles, not corpuscles this time but as packets of energy known as photons.

Currently science suggests that seeing the light is in the eyes of the beholder. If a photon is traveling through space and passes through one slit of the correct size, then the touch of consciousness will produce a particle. If a darn-near identical photon instead passes through two slits of the correct size and separation, then the touch of consciousness will produce a wave. So how can one scientist see the light as a particle and another scientist see the light as a wave? Isn't this a contradiction?[352] Can both scientists be any more correct than Buddhists and Christians making conflicting statements about the same light?

Perhaps the skeptics are right. Light is mystery; we are in the dark about the light. As we approach the infinite, the philosopher's distaste for contradiction becomes moot. Particle light, wave light, Christian light, Buddhist light—all converge through the right glass. Those who live in the light should not use it burn holes in others' stories. Even the rhetoric of objectivity fails when God, "seen through a glass, darkly," becomes as unexplainable as quantum theory. Happily, if you find such metaphors preposterous, then you may be certain you have found the true focus.

Insignificant Figures

Barring the infinite, sometimes the numbers in the rhetoric of objectivity remain impressively large. But, how many are significant? In science, there's strength in numbers—"significant figures," to be exact. Consider, for instance, calculating Red Riding Hood's speed on the way to Grandmother's house, which, according to Red, is about two miles away. The wolf's "atomic" stopwatch is accurate to a millionth of a second. He clocks Red's walk at 1.000005 hours.

Dividing distance by time, we get two miles divided by 1.000005 hours equal to—use your calculator—1.99999 miles per hour. The accuracy of the wolf's stopwatch balances the imprecision of Red's estimated distance. Right?

Wrong. As intended, the 1.99999 figure impresses us with its precision and lulls us into believing the calculations are "scientific." However, the actual accuracy of the calculation can be no more exact than its least exact measurement, the one with the lowest number of significant figures. While the Wolf's time is accurate to (count them: 1.000005) seven significant figures, Red's distance is accurate to only one significant figure—*two* miles. The accuracy, therefore, hinges not on the atomic precision of the watch but the uncertainty of Red's estimate. *1.99999 miles per hour* may look great on paper, but a simple *about two miles per hour* is as much accuracy as the significant figures support.

Rhetoric too often ignores significant figures when promoting theories. Atomic clocks, tree rings and ice cores used to date fossils, computers analyzing the genomes of Neanderthals, atomically minded microscopes reading the minds of atoms—all amaze and persuade. Still, no matter how many links are forged from the titanium and silicone of computers, clocks, or microscopes—no matter how many links are tempered with painstaking research, accumulated facts, peer-reviewed articles—the strength of our chain is not determined by its length. The strength is determined by its most fragile reconstruction, extrapolation, approximation; its frailest *if, may, might*; its boldest guess. The chain's strength as an ultimate truth claim is determined by its weakest link.

Only in rhetoric does the length of the chain matter. Don't confuse rhetoric with science. In science, rockets rocket and transistors transist according to the accuracy of the least significant figure—no *ifs, mays,* or *mights* allowed. Darwinians have faith that the gaps in the geological record are somehow more material, mightier, more holy, than the god of the gaps. Alexander Pope (1688-1744), in his poem *An Essay on Man,* described "the Chain of Being" that, before Darwin, naturalists used to bring order to the world:

> Vast Chain of Being! Which from God began,
> Natures ethereal, human, angel, man,
> Beast, bird, fish, insect, what no eye can see,
> No glass can reach! from Infinite to thee,
> From Thee to nothing.

Pope's chain stretched from "nothing" through "beast," "man," "angel," to "God." Darwin reordered the world with his chain of natural selection/organismic capitalism, adding apes and deleting angels, if not deleting God. So what is the tensile strength of natural selection? Which figures are more significant, those representing angels or evidence imagined to exist where as of yet "no glass can reach" in the imperfect geological record? Which chain is stronger, the Chain of Being or the Chain of Missing Links? The skeptic keeps watch for insignificant figures.

Rhetorical Pictures

If a rhetorical picture is worth a thousand rhetorical words, then the book *99% Ape: How Evolution Adds Up*, with its hundreds of photographs, is quite the tome. Pictures show us the human animal along such great apes as orangutan, gorilla, and chimpanzee. Pictures show us with intimate detail how our own pubic lice have evolved compared to gorilla lice. Pictures tell the story of a "tentative" ancestral tree of *H. sapiens* with *Homo habilis* and *H. heidelbergensis* and *H. neanderthalensis* going their merry evolutionary ways despite "questions that are begging for answers."[353] Pictures reveal the variation in eyes between owls and humans and spiders and insects and squids and snakes, the variation in skin pigment between Asians and Africans and Middle Easterners, Europeans, and black-hatted and white-hatted peppered moths.

The photographs are as beautiful and deceptively convincing as the composite, statistically generated pictures giving flesh and persuasive power to fMRIs. In *99% Ape* the diagrams are clear, the prose simple. The book is perfect for proselytizing to puerile youngsters and puerile adults alike about the wonders of natural selection. As always, colorful finches are highlighted as examples of how creatures evolve to fit their environments. All that is missing are the fMRIs of apes and humans to prove their commonalities and common ancestry—with Technicolor.

Primate Math

99% Ape: How Evolution Adds Up does not rely on rhetorical pictures only. It relies on rhetorical numbers: "In the ... human genome, there are 23 chromosome[s] ... Most cells of your body contains two copies of each chromosome, one copy derived from your mother, the other from your father. If you yourself have children, only one copy of each of your 23 chromosomes is transmitted to them."[354] Lest the real numbers get too boring, the book combines the numbers with an amusing anecdote:

> The third way in which sexual reproduction contributes to genetic shuffling is when a sperm fertilizes an egg, uniting the chromosomes (23 in humans) of each parent in a new alliance of chromosome pairs. Boy meets girl is the part of the sexual process with which everyone is familiar, but not everyone fully appreciates. A glamorous actress wrote to the intellectual and playwright George Bernard Shaw (1856-1950) proposing that they conceive a child, asking him to imagine how wonderful it would be for the offspring to have 'your brains and my beauty.' Shaw reputedly remarked, 'But what if he were to have your brains and my beauty?'[355]

I enjoyed the urban myth better when it was about Albert Einstein and Marilyn Monroe, but the passage is nevertheless instructive. It implies that only half your chromosomes come from your father, and your son has only half your chromosomes or one fourth of his grandfather's (your father's) chromosomes. Now, let's add this information to that found in another paragraph:

What does it mean to say that humans and apes, or more specifically chimpanzees, are 99% similar at the genetic level? How can a 1% difference between chimps and humans explain all the obvious differences between the species *Pan troglodytes* (chimpanzees) and *Homo sapiens* (us)?[356]

Once, Latin was used to sell religion; science borrowed the rhetorical device. This paragraph's Latinate *Pan troglodytes* and *Homo sapiens* assure us we are dealing with science. Tell me, though: if you share "99%" of your genetic material with chimpanzees, then how come you inherit only 25 percent of your genetic material from your father's father? Are you more closely related to Nim Chimpsky than to your grandfather? Do the numbers really add up?

Sure, I am equivocating on the term "genetic material," confusing genetic material as sequences in biochemistry with genetic material as chromosomes. My own argument is an example of rhetorical numbers; it calls into question the 99 percent figure. That is not the point. Without getting into the percentages of great ape ancestors swimming in our gene pool, visiting the simian slave pens of history, or examining the tricks in the language, the answer is actually quite simple. The statement about the grandfather is real math. The same type of math predicts that if a tall pea plant marries a short pea plant, their children will all be tall, but if these children marry each other, one-fourth will be tall, one-fourth will be short, and one-half will be arrested for incest.

Genetics is real science. It allows colorblind couples to predict their odds of producing colorblind children. The *99% Ape* math is rhetorical. Unless one is planning to raise a family with a gorilla, the math predicts only that someone is selling ideology.

To repeat the equivocation, how do we reconcile that only one-fourth of your genetic material comes from your father's father, while you share 99 percent of your genetic material with apes? *Musical definitions* and *the universal solvent* of rhetorical reason are always available to dissolve away all difficulties, but the chimp statement is calculated for a purpose. The calculation sells natural selection. Math predicts successful actions; rhetorical numbers succeed through persuasion. Newton's math predicted the motion of apples and oranges; rhetorical numbers add apples to oranges. Real math supplies know-how, while rhetorical numbers sell know-why. Real math belongs in equations, not explanations engineered to persuade.

The triumph of rhetorical numbers in the twentieth century was to add genetics to Darwin's ingenious theory to equal what became known as "the modern synthesis." The first major book of the synthesis was R. A. Fisher's *The Genetical Theory of Natural Selection*. Fisher's training was in mathematics, and he used his towering understanding of statistics to sell Darwin's theory to a new generation of disciples. According to Stephen Jay Gould, 40 percent of the book was about eugenics, about Fisher's concern that those who rise by "moral and intellectual superiority" spend less time reproducing than "the less worthy social classes."[357] To put this into numbers, a homeschooling mom teaching creation science to her ten children is about ten times more likely to pass on her genetic material than a female Ph.D. too busy studying Darwin to have more than one child. Apparently, the theory of natural selection works 90 percent better when you do not believe in it, and there is no better way to turn natural selection into a darn-near null hypothesis than to study Darwin.

Did the real genius of the modern synthesis—like my most recent argument—lie in its math, or rhetorical numbers? My homeschooling example is obviously rhetorical. Gould himself, less obviously, is using the "40 percent eugenics" figure as a rhetorical number to discredit the modern synthesis, a paradigm that Gould hopes to replace with one of his own.

Nurture Versus Nature

Another example is pertinent. Evolutionary psychologists love to preach on the extent to which our behavior is genetically determined. Outside of evolutionary psychology, not all agree. Shakespeare began the controversy when, in *The Tempest*, he penned the line, "A devil, a born devil, on whose nature/Nurture can never stick."[358] Thus, we have the question of nurture versus nature: what percentage of our makeup is due to nature, or heredity, and what percentage is due to nurture, or our environment?

Environmentalists and naturalists have been heatedly arguing the percentages ever since philosopher John Locke (1632-1704) proposed that at birth the mind is a *tabula rasa,* or blank slate. The thinking of evolutionary psychologists—and their insistence on the undue influence of heredity on human behavior—is the best evidence that not only are we born with a blank slate, but that the slate can remain blank.[359]

As we have previously calculated, although your father and your son share 99 percent of their genetic material with chimpanzees, your son inherited only 25 percent of his chromosomes from his grandfather. So what percentage of your son's behavior is inherited from his grandfather? The answer is simple. Compare your father's ability to surf the Internet when he was ten years old to your son's ability to surf the Internet at the same age. If your son's abilities were, say, a thousand percent better than your father's despite the two sharing only 25 percent of their chromosomes, then nurture, not nature, dominates. Evolutionary psychology is perhaps better suited for counting the number of fingers and toes we use while operating the mouse and keyboard or performing math without our calculators.

The above example is, of course, another example of rhetorical numbers. Throughout these essays, I have touted skepticism. What is the probability that I am correct to doubt? Consider evidence presented in Daniel Dennett's latest, *Intuition Pumps and Other Tools for Thinking.* Dennett presents his version of Sturgeon's Law (named after sci-fi writer Ted Sturgeon):

Ninety percent of everything is crap. Ninety percent of experiments in molecular biology, 90 percent of poetry, 90 percent of philosophy books [but evidently not Dennett's present book], 90 percent of peer-reviewed articles in mathematics—and so forth—is crap. Is that true? Well, maybe it's an exaggeration, but let's agree that there is a lot of mediocre work done in every field. (Some curmudgeons say it's more like 99 percent, but let's not get into that game.)[360]

No need to. Dennett has already placed it in our minds without having to defend it—just as time and again I have used rhetorical questions on your mind, without having to support the questions. Still, I like Dennett's version of the rule. It tells us that my skepticism has a 90 to 99 percent chance of being correct. When it comes to the basis of his theory, Darwin had doubts of his own. In a July 3, 1881, letter to William Graham, he wrote:

With me the horrid doubt always arises whether the convictions of man's mind, which has been developed from the mind of the lower animals, are of any value or at all trustworthy. Would any one trust in the convictions of a monkey's mind, if there are any convictions in such a mind?[361]

Darwin' doubts were justified. And so are the good skeptic's. Twenty-first century creationists and Darwinians agree with mathematical certainty: to accept that two fantasies can evolve into the science of evolutionary psychology is to accept a theory that is 99 percent ape.

16

Art, Nature, and Reconstruction

In biomythology's rhetoric of objectivity, no device is more convincing than the art of reconstruction. Let's consider, in our primer for skepticism, if we can trust the hasty generalization our search reveals.

We love to see the past reconstructed. Recently my wife and I visited the Petrified Forest National Park. The Petrified Forest Museum Association book promised an adventure:

> Step back in time and explore one of the world's greatest storehouses of knowledge about the Dawn of the Dinosaurs. Discover fossils from a time when the desert Southwest was a vast humid basin crossed by broad meandering rivers, crawling with strange creatures and covered with towering trees. [362]

The scenes from the Late Triassic ecosystem were no disappointment. They were stunning. A winding trail meandered by remarkable fossil trees from the period. The trees just happened to lie next to the trail. More remarkably, to reveal the visually delightful mixture of iron minerals and quartz, the fossil logs had been cut by diamond-hard saws, probably in some lost geological period. It was all very authentic, providing the skeptic with a real appreciation of how reconstruction works. The history of the technique goes back at least two centuries.

Dem Bones

The terror of reason and the French Revolution cleared the way for new minds. One of these belonged to the naturalist Georges-Léopold-Chrétien-Frédéric-Dagobert Cuvier (1769–1832), who, as legend has it,[363] first revealed his genius at the age of three by spelling his name. It was therefore no surprise when in 1819 he was made a peer in honor of his scientific contributions. A victory for ecology and a yet-to-bud "green" movement, the economy of the salutation *Baron Cuvier* shortened the proceedings of the French Academy by months and saved rivers of ink and forests of trees. A commensurate saving of resources would not be seen again until Darwin's *On the Origin of Species by Means of Natural Selection, or the Preservation of Favoured Races in the Struggle for Life* was expurgated to the politically correct *The Origin of Species* (not that Darwin was necessarily writing of human "races" any more than racism was politically incorrect in Darwin's Victorian England).

Cuvier, a self-styled expert in comparative anatomy, studied the bones of Indian and African elephants and compared them to the fossil remains of northern mammoths. He concluded that the mammoths were extinct, destroyed by a catastrophe. His lectures about extinctions and catastrophes captivated the public and secured his fame. Finding few gradual changes in the actual fossil record, Cuvier ridiculed his senior colleague Jean-Baptiste Lamarck—in Cuvier's mind a dangerous fossil from prerevolutionary days—for a theory of transformation (evolution) more based on speculation than direct empirical evidence.

Lamarck, Evolution, and Paupers' Graves

Lamarck postulated that organisms had transformed over time by a "'power of life,' a characteristic possessed by living things that made them increase in complexity over time."[364] This change for increasing complexity came from within individuals, use and disuse of body parts helping to shape the change. The most famous example of this was the necks of giraffes hypothesized to have grown longer from stretching to eat leaves on trees. Lamarck proposed that these acquired characteristics could be inherited by subsequent generations. Darwin, hedging his bet, included Lamarck's "use and disuse" in the final paragraph of *Origin*. Epigeneticists are currently exploring the possibility that acquired characteristics may be inherited by altering gene expression, thus helping to validate Lamarck. In Lamarck's day, however, Cuvier lampooned the concept, claiming that "hens searching for their food at the water's edge, and striving not to get their thighs wet, succeeded so well in elongating their legs that they became herons or storks."[365] Possibly responding to Cuvier's ridicule, naturalists more often spoke about Lamarck behind his back than they spoke about his theory in public.

Lamarck's reputation dwindled. Eventually, he was, as previously mentioned, buried in a pauper's grave.

Catastrophe Versus Gradualism

What goes around comes around. Within a year of Cuvier's death, his theories were already being attacked. In 1833, the geologist Charles Lyell wrote of Cuvier's world of fits, starts, and catastrophes, "Never was there a dogma more calculated to foster indolence, and to blunt the keen edge of curiosity, than this assumption of the discordance between the former and existing causes of change."[366] As is often the case, the live Lyell won the debate against the dead Cuvier—for a time, at least—and catastrophes were whisked aside to make way for the gradualism upon which Darwin's natural selection rests. Until being resurrected by Gould in 2002, Cuvier continued to be ridiculed and completely misinterpreted:

> Cuvier believed that Noah's flood was universal and had prepared the earth for its present inhabitants. The Church was happy to have the support of such an eminent scientist, and there is no doubt that Cuvier's great reputation delayed the acceptance of the more accurate view that ultimately prevailed.[367]

This was written in 1973 and used to summarily dismiss Cuvier because of his supposed creationist sympathies. Since then, the theory of the catastrophe of a comet hitting the earth and causing a mass extinction has been voted in by consensus. The truth of science has, once more, changed, and the "more accurate

David Cook

view" has once more prevailed. Science just keeps getting better and better, even though it was evidently worse for a time between Cuvier and us.

Correlating the Parts

For the purposes of the present essay, Cuvier left a far greater legacy than catastrophism. He armed the rhetoric of Darwinians with "the principle of the correlation of parts," sometimes recounted as, "Toe bone connected to the foot bone; foot bone connected to the leg bone." As mentioned, Cuvier was a comparative anatomist. Having won his fame on the comparative anatomy of living elephants and extinct mammoths, Cuvier taught that the parts of a skeleton are related, so that—if you happened to share Cuvier's genius—you could infer the whole of an organism from any part—a goddess, perhaps, from a toenail clipping?

Reconstruction

Paleontologists use this principle in *reconstruction*, constructing a whole body from a shattered bone or two, extrapolating soft eye color from hard eyeteeth. One of Cuvier's followers who was particularly adept at reconstruction was Sir Arthur Smith Woodward, whose career was described by fellow knight Sir Arthur Keith in the forward to Woodward's 1948 book:

> At the age of eighteen, he [Woodward] became an assistant in the Geological Department of the British Museum (South Kensington) and there he labored for

forty-one years; he had been head of his Department for twenty-two years before his retirement in 1923, at the age of fifty-nine. His career was crowded with one discovery after another; how full these years were is vividly illustrated by the fact that the mere list of addition he made to the knowledge of his time occupies twenty-four pages of The Proceedings of the Royal Society.[368]

Keith was hardly exaggerating about Woodward's scientific stature. The former Keeper of Geology at the Natural History Museum, President of the Geology Society, and Fellow of the Royal Society also won the Queen's Medal of the Royal Society. The medal had previously been given to the likes of John Dalton, Michael Faraday, and Charles Darwin for the most important contributions to the advancement of natural knowledge. Not only was Woodward a scientist of scientists, he was an expert at reconstruction: he reconstructed an entire primitive man from two teeth, a jawbone, and a piece of a shattered skull. Indeed, an illustration of the man, with his broad nose, sloping forehead, shaggy hair, and hairy face and back appears on the title page of Sir Arthur's book, *The Earliest Englishman*.

The earliest Englishman's story began in the days before the World Wars, when the sun never set on the British Empire. At that time, it was only natural that the soil of England should house the truth of Man's descent. Therefore, it was that the charming Southeastern English village of Piltdown offered up a fossil skull with a human braincase, two human teeth, and an ape-like jawbone.

David Cook

Piltdown Man

Thus was born "Piltdown Man." The braincase, first mistaken for a coconut and shattered with a shovel by gravel-pit workers at Barkham Manor Farm, was discovered by Charles Dawson, an amateur fossil collector with an "inquiring mind." The piece of shattered human skull brought to Dawson by the workers was of "unusual thickness and texture." Dawson eventually enlisted Woodward's aid. By 1912, what they reconstructed from the gravel pit elevated England to its rightful lead in the world of evolution. Woodward fleshed out the Piltdown bones (and teeth) by meticulous anatomical examination, by X-ray analysis, by tracing the geological history of the gravel to the *Pleistocene*, by providing tables of *Geological Ages* and photographs of the skull and jaw fragments and illustrations comparing ape and human anatomy, by logically winnowing the manmade from the natural, and by careful examination of other fossils in the gravel. In short, he used the tools of scientific empiricism to connect the dots of history.

Following Your Paleontological Bliss

As mentioned, Woodward's discovery was notable because it was found on English soil. In 1891, another missing link had been found on the island of Java. The story of the discovery, though somewhat of a sidetrack, is worth telling. Eugene Dubois (1858-1940) was a paleoanthropologist who as a boy dreamed of finding "the missing link" between ape and man. Dubois forsook his medical career and his European academic career to pursue his dream. He joined the Dutch army for the purposes of traveling to what is now Indonesia, where he believed he would find his prize. He discovered the missing link, "Java Man," in 1891, two decades before Piltdown Man. The story is interesting because it inspires us to ask the question: do our dreams guide science, or does science confirm our dreams? Who can say? Since Dubois' day, Java Man—no Englishman— has been renamed *Homo erectus* and shuffled off to a side branch of the hominid family tree. He no longer falls on the direct path between ape and man. Although he continued to keep hope, and digging, alive, he was a missing link that missed its link, so to speak.

Another boy who dreamed of finding the missing link was Donald Johanson. His book, *Lucy: The Beginnings of Humankind*, tells the story:

> Never in my wildest fantasies did I imagine that I would discover a fossil as earthshaking as Lucy. When I was a teenager, I dreamed of traveling to Africa and finding a 'missing link.' Lucy is that and more; a 3.2-million-years-old skeleton who has become the spokeswoman for human evolution. She is perhaps the best known and most studied fossil hominid of the twentieth century, the benchmark by which other discoveries of human ancestors are judged.[369]

In the words of Jesus (almost), "Seek a missing link and you shall find it." Therefore, we have a gold standard in Lucy, named after "Lucy in the Sky with Diamonds," the Beatles tune playing as the old girl's discovery was celebrated. Gold and diamonds—the story

David Cook

provides a rich heritage, a certain certainty. Even today, though, the branches of the hominid family tree range anywhere between uncertain, tentative, and fairly certain. Tales of reconstruction, like Lucy's, nevertheless enjoy the "benchmark" of certainty. One wonders if the story is proof of our hominid ancestry or a testament to Rhonda Byrnes' *The Secret* and her "law of attraction": keep your mind's eye on the diamond mine, and you'll attract it into your life! Or imagine you have, which is just as good.

We find the same certainty in these quotes from Woodward's *The Earliest Englishman*: "He [Piltdown Man] certainly was a man, and not a creature half-way between man and ape. He was perhaps ungainly, and may have walked with a shuffling gait, but his brain and skull were essentially human."[370] Woodward's cautious attention to empiricism is highlighted in the following paragraph:

> When the brain-cavity of the final reconstruction of the skull was measured, it was found to be much larger than the smallest known brain-cavity of modern man. It proved to be more than twice as large as the brain-cavity of the largest known gorilla ... Although it cannot be measured exactly, it seems to have contained a brain at least 1,350 cubic centimetres in bulk. The shape of the brain is typically human, but according to Elliot Smith it is incompletely expanded ... The absence of this growth shows that Piltdown Man was adapted only for a simpler life than modern groups of men. He could doubtless speak, for the width of his jaw would allow him to use the tongue freely for forming definite sounds. One of the least expanded parts in the brain, however, is that which governs reasoning and the association of ideas, so that his powers of speech

must have been limited. Many of the folds in the structure of the brain have left clear impressions on the bony wall of the brain-cavity, and Keith has observed that some of these suggest an arrangement more ape-like than any seen in modern man.[371]

What could be more scientific? Woodward, continuing the paragraph, also mentions "the late Professor J. Symington" bringing up difficulties. Woodward concludes, wondering "whether the late Sir Grafton Elliot Smith felt these difficulties when he had presented his detailed account of the brain of Piltdown Man to the Royal Society in 1914, because he never completed the paper for publication."

Woodward could have learned the art of name-dropping from Darwin himself, mentioning in a single paragraph Keith, Professor Symington, and Sir Grafton Elliot Smith, although Darwin would surely have used Keith's full name, Sir Arthur Keith. Woodward was similarly expert at rhetorical empiricism; he gave us the exact cubic centimeters of the brain and the carefully anatomically reasoned data on Piltdown Man's powers of speech to shore up his argument with facts. Woodward demonstrates his keen reasoning ability when dismissing the doubters (probably French) of the match between the jawbone and the skull:

> The lower jaw is in several ways so remarkably ape-like that some have doubted whether it really belongs to the human skull which was found near to it. They have supposed that at Piltdown we have discovered an entirely new kin of human skull lacking its lower jaw, and an equally new kind of ape jaw lacking the skull. This would be a startling result

indeed to achieve in a single cubic yard of gravel. As the lower jaw contains two grinding teeth which are essentially human, it is more reasonable to conclude that the new skull and the new lower jaw belonged to the same head.[372]

Based on exhaustive scientific observation and reasoning, Woodward eventually concluded: "The human skull of Piltdown is ... one of the oldest discovered anywhere in the world."[373] Then, "[The skull] of Piltdown Man is chiefly remarkable for its ape-like lower jaw. Fossils leave no doubt indeed, that man is descended from ape-like ancestors to whom he owes most of his bodily shape."[374] In addition, "It may now be observed that the Piltdown skull forms a direct link between the supposed earliest ancestral apes and modern man."[375]

In gratitude, Woodward christened Piltdown Man with the name *Eoanthropus Dawson* to credit the fossil's discoverer, and "from two Greek words that mean Dawn-man."[376] Sir Arthur Keith—who was once gracious enough to unveil Mr. Dawson's stone pillar memorial at Barkham Manor—supported his fellow knight's enthusiasm in the book's foreword. "I am firmly convinced that no theory of human evolution can be regarded as satisfactory unless the revelations of Piltdown are taken into account."[377]

How the Glitch Stole Christmas

Amen. I certainly agree: *no theory of human evolution can be regarded as satisfactory unless the revelations of Piltdown are taken into account*. There was, however, one glitch in the tale.

In the October 1955 issue of *American Scientist*, Oakley and Weiner detail how Piltdown man was a hoax: a modern human braincase, the jawbone of an orangutan, the teeth of a chimp. The teeth had been filed to appear human, the bones stained to match the gravel and fractured to help conceal the fraud. All of Woodward's scientific tables, facts, observations, and reasoning could not put Piltdown Man back together again, although half a century for a scientific fact isn't all that bad.

Who committed the fraud? The answer is lost somewhere in the imperfect fossil record of scientific history. Is the tale notable because it cautions us against scientific fraud? No. Woodward was a consummate scientist and completely sincere. Indeed, he would have been brokenhearted had he lived long enough to witness the end of the tale. It really doesn't matter that he was the victim of a practical joke because Mother Nature herself is also all too fond of practical jokes. Every coincidence, every anomaly, every monster is a practical joke of nature, ripe for serving as a character in a scientific story.

The Piltdown tale exemplifies how science actually works. Science is the search for truth. The Rhetoric of Objectivity is used to sell us on what to believe. Passions and personalities—and in this

311

case English chauvinism—eventually agree on which perception of the evidence to teach to schoolchildren. In the end, a new scientific discovery detects the error. The moral: as we have emphasized in a past chapter, science may be the road to truth, but until the roadwork is done, the evidence complete, the last of nature's facts uncovered, don't confuse the road with the destination.

Nature Imitates Art

Oscar Wilde, in his 1889 essay "The Decay of Lying," insisted that "Life imitates Art far more than Art imitates Life." When it comes to the reconstruction of fossils, he may well be right. Since Woodward's Piltdown fiasco, scientists have learned the error of their ways. When reconstructing, they now defer to computer modeling to authenticate the figures populating our museum exhibits. Although, can technology really balance "educated" guesses? Is extrapolation a euphemism for imagination? Do reconstructions reflect nature or art?

A recent visit to a major museum of natural history provided me with evidence for answering these questions more intelligently and truthfully, eliminating my need to substitute reason for unmitigated passion.

In the "Hall of Human Origins," I found a group of primitive hominids (humans or human ancestors or something resembling them) huddled around a fire for warmth. From the noble savages and savagettes dangled flaccid penises and breasts. The detail, however, that really sold me—salvaged from a swath of natural selection's wreckage in the imperfect fossil record—was the perfection of the manicures and pedicures! No chipped claws, torn cuticles, or unsightly hangnails here—humans, having longer legs, were apparently able to walk further than apes to have their nails done. That the nails share a dull base coat and no glossy top coat precisely captures the evolutionary stage of our ancestral nail technology and proves the power of computer modeling in revealing truth.

When schoolchildren visit these exhibits to "consider the evidence for themselves," as Richard Dawkins implores them to do, don't let them confuse products of nature with products of the human mind or with the boyhood dreams of paleoanthropologists proving the law of attraction. Demand actual fossil remains, preferably arranged as they were discovered, so we can separate the nature from the biomythology. As every good skeptic should consider, unless "natural history" is a synonym for "artistic license," the reconstructions belong in a good museum—not of natural history, but of modern art.

17

The Ghost in the Paradigm

No primer on skepticism can boast a complete survey on the rhetoric of objectivity without visiting the brainchild of Thomas Kuhn (1922–1996). Kuhn was a philosopher and historian of science whom skeptics love to love and realists love to hate. In *The Structure of Scientific Revolutions*, Kuhn peeked at the connect-the-dot books of science. He found that scientists loved to compete in connecting the dots of fact, previously numbered by genius, to create pictures of nature. In the process, scientists would use the crayons of experimentation, observation, and reason. Unnumbered factual dots would often begin to materialize, ruining the aesthetics of the existing picture.

Scientific genius is the ability to see past belief. When a picture became too confused, a new creative genius might step back to reimagine nature, renumber the dots, and create an astonishing and alluring new picture. While the older dot-connecters—convinced by the apparent cause and effect of the habitual actions of a life's work—might cling to the old picture and reject the new, the younger generation of dot-connecters would consider the new creation. Providing that the picture required fewer—or at least no more—bent lines to conform to observation and providing the picture was simple, elegant, beautiful, and brilliant, the younger dot-connectors would agree with the new art of nature and begin again in earnest using their crayons of experimentation, observation, and reason to compete in connecting the dots according to the updated number scheme.

The connecting-the-dots metaphor is mine. Kuhn preferred "puzzles" to "pictures." These puzzles he called *paradigms*: "universally recognized scientific achievements that for a time provide model problems and solutions to a community of practitioners." What I called, connecting the dots, Kuhn called *puzzle solving* or *normal science*, the "research firmly based upon one or more past scientific achievements, achievements that some particular scientific community acknowledges for a time as supplying the foundation for its further practice."[378] Paradigms are the favorite pictures and puzzles voted in for adoption during a particular era. They come and go. While one remains, it focuses attention like a magnifying glass, burning sizable holes in mystery even while it blinds us to all that falls outside the optical limits of its glassy-eyed focus.

Why do scientists embrace new dot-to-dot drawings, new puzzles? Kuhn wrote:

> When ... the profession can no longer evade anomalies that subvert the existing tradition of scientific practice—then begin the extraordinary investigations that lead the profession at last to a new set of commitments, a new basis for the practice of science. The extraordinary episodes in which that

315

shift of professional commitments occurs are the ones known in this essay as scientific revolutions.[379]

According to Kuhn, "Competition between segments of the scientific community is the only historical process that ever actually results in the rejection of one previously accepted theory or the adoption of another."[380] Thus, Kuhn sees the winning pictures or puzzles, the winning paradigms, being dictated not just by their fit with nature but by competitions between the scientists themselves.

Only a Paradigm

As described, in Cobb County, Georgia, where I live, the Board of Education applied a sticker inside its science textbooks, "Evolution is a theory, not a fact, regarding the origin of living things."

In his finer rhetorical moments, Darwin might have agreed, "There is a grandeur in this view of life, with its several powers, having been originally breathed into a few forms or into one." Darwin knew that natural selection was a *view* of life, a life seen from one perspective. For Cobb County, it was "a theory, not a fact." For Jerry Coyne, it was *Why Evolution is True*. For Richard Dawkins, it belonged to a "list of certain facts that will never be disproved."[381] For Kenneth Miller, it was "the central organizing principle of the biological sciences." Now, while it is true that Coyne, Dawkins, and Miller are, in words, speaking about evolution, not natural selection, this is mere equivocation. As the skeptic has seen, the titles are about evolution, but the books are about natural selection.

Evolution by natural selection is more than a theory; it is the direction-setting agreement among scientists during a particular period of history. It is a paradigm.

Paradigms come. Paradigms go. Consider a quote from a letter that Alfred Russel Wallace wrote to his friend George Silk, September 1, 1860: "I have read it [Darwin's *Origin*] through 5 or 6 times & each time with increasing admiration. It is the 'Principia' of Natural History. It will live as long as the 'Principia' of Newton."[382] Wallace's analogy was prophetic. At the hands of Einstein's new paradigm of relativity, the absolute time, space, and matter of Newton's *Principia* would evaporate within fifty years of Wallace's letter. For 200 years, Newton had been no less admired than Darwin is now. Newton's paradigm explaining motion on earth and in the heavens was the most renowned to come along since Aristotle's. Newtonian mechanics, with its prediction and control of the physical world, inspired the well-earned prestige that real science would come to enjoy.

The skeptic must wonder. Is there a reason to suppose that the stretch of Darwin's story will prove more elastic than Newton's? Can dot-to-dot drawings fully capture nature? Is reality really dappled? Is natural selection true? To the mind of the skeptic, the probability is that the numbering of nature's dots will continue to evolve to fit the needs and pretensions of passing cultures hoping to establish the superiority of their superstitions compared to those of their unenlightened ancestors.

Paradigms Evolve

Kuhn has looked at paradigms as the evolution of scientific thought and asked if there were any reason to suppose that thought was advancing rather than just changing. As you will remember, Darwin's theory shuns teleology. Evolution has no specified goal, no purpose, no necessary direction. Only change—for better or for worse, so long as that change fits the organism or theory to the environment or culture at hand. From this perspective, Kuhn questions if there is any reason to suppose that evolution of paradigms provides progress, "[a]nd why should paradigm change invariably produce an instrument more perfect in any sense than those known before?"[383]

In other words, are the fashions in science becoming ever more perfect, or are they merely evolving? The skeptic could wonder: was banning vital forces from life a step forward? If our thinking is dictated by our evolution, are we confusing "progress" with adaptations and conformations to shifts in the intellectual, cultural, and political environment at hand? Are the flashes of creative genius behind paradigm changes like random comets reconfiguring thought on earth? On the other hand, as Thomas Nagel has suggested, is there a method to the madness of nature, a method tied to consciousness and value?

Mind Over Matter

In a more recent adaptation of science to culture, the paradigm of the ethereal soul has been replaced by the paradigm of the gelatinous brain. This paradigm shift should provide a summary and conclusion to this chapter on how scientific tastes change with time to conform to the passing fancies of history.

In 1994, two Nobel laureates published popular-science books on neuroscience. Sir John Eccles (1903-1997), who had been trained by Sherrington and had already devoted sixty years of his life to the subject, wrote *How the Self Controls Its Brain*. He began with a quote headed "Shakespearean Dualism":

> My brain, I'll prove the female to my soul,
> My soul, the father; and these two beget
> A generation of still-breeding thoughts;
> And these same thoughts, people this world
> In humours, like the people of this world,
> For no thought is contented.
> William Shakespeare, *Richard II*, Act V, Scene 5

Eccles substituted the more politically correct *self* for *soul* but argued that Shakespeare's older view (and the view of most of humanity) was still essentially correct, that the self controlled its brain to carry out countless planned, voluntary actions each day. There was, however, a movement afoot to squelch such thinking. As Eccles told the story:

> Promissory materialism is simply a superstition held
> by dogmatic materialists. It has all the feature of a
> Messianic prophecy, with the promise of a future

freed of all problems—a kind of Nirvana for our unfortunate successors. In contrast the true scientific attitude is that scientific problems are unending in providing challenges to attain an even wider and deeper understanding of nature and man.[384]

Having spent sixty years on the subject, Eccles was a real authority. He may have fought the good fight, but he lost the battle. The profession followed the rallying cry of Francis Crick—the second laureate publishing in 1994 on neuroscience. Crick was a physics major turned molecular biologist who, along with James Watson, had borrowed the X-ray work of Rosalind Franklin. Watson and Crick met the Nobel committee. Rosalind—transported by ovarian cancer, rather than a nod from the Nobel committee, for the time, at least—met her maker.

Ironies aside, Crick could make a lasting impression. James Watson opened his memoir with words about his co-discoverer of the structure of DNA:

> I have never seen Francis Crick in a modest mood. Perhaps in other company he is that way, but I have never had reason to so judge him. It has nothing to do with his present fame. Already he is much talked about, usually with reverence, and someday he may be considered in the category of Rutherford or Bohr. But this was not true when, in the fall of 1951, I came to the Cavendish Laboratory of Cambridge University ... Although some of his colleagues realized the value of his quick, penetrating mind ... most people thought he talked too much.[385]

Watson was as much a master of understatement as Crick was a master of overstatement. With characteristic aplomb, Crick went on to propose "the central dogma of molecular biology." According to the dogma, the traffic of information from DNA to protein flows down a one-way street. The ill-conceived and fated dogma shortly died of AIDS, one of the retroviruses found to drive against Crick's proposed traffic flow. That the "central dogma" would soon decompose into dog food didn't bother Crick. As he had earlier abandoned physics for biochemistry, he now abandoned biochemistry for neuroscience. This time, instead of dogma, Crick modestly offered *The Astonishing Hypothesis*:

> The Astonishing Hypothesis is that 'You,' your joys and sorrows, your memories and your ambitions, your sense of personal identity and free will, are in fact no more than the behavior of a vast assembly of nerve cells and their associated molecules. As Lewis Carroll's Alice might have phrased it: 'You're nothing but a pack of neurons.'[386]

Since Crick was not writing to advance the science of the mind, but to change the minds of the lay public, he explicated the passage with a footnote: "'Neuron' is the scientific word for a nerve cell."[387] *Scientific*? At least we are spared the word *consensus*.

Matter over Mind

The skeptic could imagine that back before Cupid had traded in his bow and arrows for an fMRI, love had been brainless; now it is heartless. However, unlike "loving with all our hearts," "thinking with our whole brains" had escaped the province of metaphor and elevated to science. Crick's cheery prescription cured neuroscience of dualism and any fantasy that a human being may transcend the tools of science; it cleared the way for a new paradigm. In the words of Thomas Kuhn, "the man who continues to resist after his whole profession has been converted has *ipso facto* ceased to be a scientist." To remain a dualist like Eccles is no longer to be a scientist. The argument from ignorance—"We do not understand how the non-material could affect the material"—has prevailed. Only the discovery and sale of a new analogy, metaphor, or myth could kill it.

Offal Reasoning

The skeptic may wonder: who sits closer to truth, the sage blind to mystery or the fool blind to all else? As Christian apostate Michael Shermer writes in one of his many books using science to persuade the lay public to join him in his apostasy, "These studies, and countless others, continue to rain blows down upon the dualist head that the brain and mind are separate. They are not. They are one and the same thing."[388]

In *Cosmos*, Carl Sagan tells us, "We are made of starstuff." Note well that Shermer is not saying that our thoughts are the inner chatter of starstuff; he is saying the chatter and brain are equal to the same stuff. Shermer is not saying that the mind emerges from the brain and, as such, is non-reducible, more than the sum of its brain parts. For the purpose of persuasion, Shermer's words are promoting an older view in philosophy of mind known as "identity theory," the notion that the mind and the brain are the same animal, kind of like *water* and H_2O or *the morning star* and *the evening star* or Batman and Bruce Wayne. Sometimes the theory is presented as, "Mental states are equal to brain states."

The skeptic should perform a simple experiment to test Shermer's idea that the brain and the mind are "one and the same thing":

> 1) Hold a quarter between your thumb and pointer finger so the profile of George Washington and the coin's date and the word *Liberty* are facing you. Extend your arm outward until the coin face is about twenty inches from your eyes. How big does the coin look?

> 2) Move the coin to ten inches in front of your nose. Look again. Does the coin, whether easier or harder to see, still look and feel to be about the same size?

In reality, the quarter is about an inch in diameter. Since this is an experiment and we are being *scientific*, we may flaunt our significant figures and allow that the quarter is .995 inches in diameter. Having performed this experiment with optometric audiences across the nation, I can tell you that, almost without exception, the quarter is said to remain the same size. In other words, those who are fiercely attending to George's profile and find themselves as perceptually blind to the rest of the world as half the spectators are blind to gorillas passing through basketball games,[389] have continued to enjoy the identical mental event of a .995 inch quarter in both positions.

How about brain states? The light reflecting from the quarter passed through the ten or twenty inches of space to be bent by the front surface of the eye (cornea) and the lens inside the eye to focus upside down on the back of the eye (retina). Interestingly, the upside-down image of the quarter on the retina was twice as large at ten inches as it was at twenty—or so the science of optics demands. Because of this change in retinal image size, the two different images were represented by different cells in the brain, by different brain events. Thus, during the experiment you enjoyed two different George Washington brain states for a single George Washington mental state of a .955-inch quarter. If the brain states are different and the mental states are the same, how can Shermer's declaration that the brain and the mind are "one and the same thing" be true? The skeptic is hard-pressed to defend mental states equaling offal states—not that the metaphor, as any skeptic could tell you, isn't close enough for science when peddling worldviews to the public.[390]

Psychologists use the terms *size constancy* and *object constancy* to explain away the fact that physical events in the retina do not account for how objects viewed from different distances and angles remain perceptually the same. The skeptic understands that size constancy explains little; it is merely another example of Huxley's declaration, "In the ultimate analysis everything is incomprehensible, and the whole object of science is simply to reduce the fundamental incomprehensibilities to the smallest possible number." Size constancy is one such reduction, used to comfort those who try to equate nervous systems to the self—even though the self may well be composed of all it possesses, including the body and the stories it tells about the surrounding environment. Retinal/brain events do not equal perceptual events, if the same mental event can be spawned by any number of different brain events, which do not equal one another. Or so the skeptic may reason.

Attend the Tale of Phineas Gage

Just as the skeptic may doubt the brain-events-are-equal-to-mind-events story, there is another popular tale of the new paradigm that invites the skeptic's doubt. Phineas Gage, the story goes, was in the right place at the wrong time when an explosion caused a three-foot iron rod to enter the left side of his face, pass behind his left eye, bore a hole through his left frontal lobe and pass out the top of his head. Before the accident Gage was, by all accounts, a polite enough fellow. Afterwards, his manners deteriorated.

I don't know about you, but I get a little cranky when I'm late for supper. I can't imagine how I would act were I to lose an eye and have my brain splatter out an open flap of my skull. Even so, according to biomythologists, Gage's behavior is proof positive that the seat of good manners is the frontal lobe. Sadly, the skeptic laments, no one saved Emily Post's brain, to examine alongside Einstein's, so we have no way to confirm the relative Herculean dimensions of her lobes.

To the skeptic, the confessions of brains obtained under the duress of neurological injury have always been at least as suspect as confessions of heresy and witchcraft obtained under torture by the inquisition. The story generally goes that such-and-such an area was damaged and that such-and-such a function was lost. Using such logic, the skeptic could as easily argue for the metaphor of the heart being the seat of love because when a three-foot rod passes through the heart all infatuations and affections abruptly cease.

That a given brain area may be necessary for a function (until another area has the nerve to take over) in no way suggests to the skeptic that the area is itself sufficient for that function without the aid of other brain, body, or environments factors. Finding a solitary frontal lobe with good manners, the skeptic could imagine, is about as likely as finding a solitary thumb hitchhiking along the road to truth. The deluge of biomythological reports that this brain module is responsible for this or that function, the skeptic might suppose, is the stuff of biomythology, not real science. Each brain area demands for its function to be connected to the rest of the brain, which is connected to the rest of the person, which is connected to the rest of us and the rest of the world. Area, brain, person, us, world—all may not be either necessary or sufficient, but all are responsible.

Pulitzer Prize winner Marilynne Robinson is similarly skeptical of the Gage tale. She makes some of the same objections, although more politely (she has larger frontal lobes, I imagine):

> I trouble the dust of poor Phineas Gage only to make the point that in these recountings of his afflictions there is no sense at all that he was a human being who thought and felt, a man with a singular and terrible fate. In the absence of an acknowledgment of his subjectivity, his reaction to this disaster is treated as indicating damage to the cerebral machinery, not to his prospects, or his faith, or his self-love. It is as if in telling the tale the writers participate in the absence of compassionate imagination, of benevolence, that they posit for their kind.[391]

This "absence of compassionate imagination" about the "subjectivity" of other minds is telling. The skeptic may doubt if an examination of Phineas Gage's frontal lobes brings us no closer to understanding what it means to be human, and certainly no closer to acting that way.

New Metaphor—Same Mystery

The philosopher of neuroscience Patricia Churchland enjoys personifying the brain. She writes, "The weight of evidence now implies that it is the brain, rather than some non-physical stuff, that feels, thinks, decides."[392] On the other lobe, contemporary philosopher Alva Noë responds in *Out of Our Heads: Why You Are Not Your Brain, and Other Lessons from the Biology of Consciousness*:

> But what needs to be kept clearly in focus is that the neuroscientists, in updating the traditional

conception of ourselves in this way, have really only succeeded in replacing one mystery with another. At present, we have no better understanding of how "a vast assembly of nerve cells and their associated molecules" might give rise to consciousness than we understand how supernatural soul stuff might do the trick. Which is just to say that the you-are-your-brain idea is not so much a working hypothesis as it is the placeholder of one.[393]

The skeptic may ask, is a life unexamined by an fMRI worth living? Times change; scientific consensus changes; paradigms come and go quicker than a blink in deep time. But, at least for now, biomythology shows no sign of abating. Put another way: mysteries outlive metaphors. While "pack of neurons" may be poetic, it will likely prove but a passing fashion in the quest to discover what makes us human. As the good skeptic intuits, scientific disciplines in their infancy make large, not small errors. Copernicus placed our sun at the center of the universe. Later, nebulae became clouds of cosmic dust; this year they are galaxies. So is that smudging on the fMRI just a smudge, a nebula, or a galaxy? Until we have the answer, the skeptic may safely assume, it is biomythology to suppose that neuroscience has the answers to what makes us human.

18

Free Will versus Free Excuse

For the past number of chapters in our primer for skepticism, we have visited devices artfully employed in the rhetoric of objectivity. Now we will examine a few ways in which applied biomythology has revolutionized our culture, inviting the skeptic to assess how "science" has been borrowed to update and solve an age-old philosophical debate: do packs of neurons have free will?

The Perpetual Argument

Philosopher Robert Kane, author of *A Contemporary Introduction to Free Will* and editor of *The Oxford Handbook of Free Will*, begins both volumes with a quote from the twelfth-century Persian poet and Islamic mystic Jalalu'ddin Rumi, "There is a disputation that will continue till mankind is raised from the dead, between the necessitarians[394] and the partisans of free will." Rumi was, perhaps, overly optimistic about the speed of the resolution. Six hundred years later, Hume was still complaining about the difficulties of reconciling liberty with necessity: "From this circumstance alone, that a controversy has been long kept on foot, and remains still undecided, we may presume, that there is some ambiguity in the expression, and that the disputants affix different ideas to the terms employed in the controversy."[395]

Kant also fears Rumi's point, distinguishing the free-will-versus-necessity problem as a "difficult problem which millennia have sought in vain."[396] Kane admits, "The problem of free will and necessity (determinism) ... is one of the most difficult and 'perhaps the most voluminously debated of all philosophical problems,' according to a recent history of philosophy."[397] In the words of American philosopher Roderick Chisholm (1916–1999), "Perhaps it is needless to remark that, in all likelihood, it is impossible to say anything significant about this ancient problem that has not been said before."[398] In this spirit of futility, we enter the fray, telling a story of our own.

Mystical Determinism

If a mystic tells us, "There are no coincidences; everything happens for a reason," we call it superstition. If a biomythologist tells us, "There are no coincidences; everything happens from its cause," we call it determinism.

Determinism commands that what comes before determines what comes next. It derives from the dogma that every effect has a cause in the past. The birth of twins, for instance, is correlated, the birth of the second twin generally following on the heels of the first. Determinism can be called upon to explain how the first twin causes the second. Determinism, the skeptic might agree, can be called on to explain just about anything, for if one thing leads to another our actions are *necessitated* or *determined*.

Experience Versus Theory

Not all agree. In the words of a man of letters who may well have given "man of letters" its name, the lexicographer Samuel Johnson (1709-1784), "All theory is against free will; all experience for it." Johnson has captured the essence of the argument. In the action world of *immediate*, habitual, perceptual experience of action, we know that billiard balls are determined and we are not, that we are free to take the shot or drop the cue. In the story world of thought, all things are possible; logic is never more than a fresh definition away from making the sale. At its heart, the free will argument is about whether or not to allow the universal solvent of reason to wash away all conflicting experiences. If reason and experience contradict each other, then one must be false. But which one?

Terminology

Those who strongly argue for freedom of will are known as libertarians (not to be confused with members of the political sect claiming that charity becomes slavery when legislated). Those who argue for what I call *freedom of excuse* are known as determinists. In his famous essay on the subject, "Freedom and Resentment," P. F. Strawson calls those who think that free will is incompatible with determinism *pessimists* and those who think that free will and determinism are compatible *optimists*. Similarly, Ivy League philosopher Harry Frankfort in his equally famous "Freedom of the Will and the Concept of a Person" distinguishes *persons* from *wantons*, with persons having volitional desires to overcome their basic desires while wantons go with the flow of their own juices.

The terms of Strawson and Frankfurt, while falling shy of outright deprecation, certainly hinge on the judgmental. This, I believe, sets a healthy precedent. Following their lead, I will, for the benefit of the skeptics, call those who favor free will *fools* (because free will sets us up to do some pretty foolish things) and those who favor determinism *idiots* (because they typically have more idiotic excuses than Marilyn Monroe had pheromones).

Defining Free Will Versus Determinism

As Hume suggested, the real problem is definition. Therefore, let's begin with definition. *Free will* is what directs our actions—until we get caught. Afterwards, *determinism* is what directs our excuses. Determinism is a euphemism for *free excuse*. In other words, the experience of free will necessitates free excuse. When freedom of will leads, freedom of excuse is never far behind.

An Excuse for a Thought Experiment

By definition, you are a fool if you think you have free will and an idiot if you think you have free excuse. Consider the following thought experiment designed for our skeptical audience:

On Speeding

A motorist, being ticketed for speeding, makes the following plea:

"I'm right sorry, officer. I was late for my grandmother's funeral (a decade late, if truth be told). My speedometer was broken, not that I noticed the darn thing was jammed at fifty-two until I reached that speed. Still, I reckon it don't matter none. We ain't morally responsible if we couldn't have done nothing else—however much we enjoyed doing what we were doing—ain't that right, officer? As for going fifty-two, I gotta confess: I am dyslexic. I reversed twenty-five and came up with fifty-two. I'm not to blame if I saw backwards. To make matters even worse, my wife, she was a-screaming in my ear, which distracted me from the speedometer big-time and reminded me of my father. He always speeded and screamed when he indoctrinated me to speed and, believe me, the acorn, it don't fall far from the tree. I'm a victim of nature and nurture. And speakin' of both, my dying mother, she just called and ordered me to visit her one last time at the hospital. I was a victim, officer, a victim of compulsion. I was being coerced by a son's love.

"To make things even worse, I have this here peanut allergy. It drives me insane. There was no warning on the peanut bag anymore this time than the last time when my windpipe swelled shut. It was just like those jerks at that fast-food chain failing to warn the poor woman that the coffee was hot. Us common folk can't be expected to know nothing without a label, now can we? Due to my allergy, my IQ dropped until I was too stupid to be held accountable. In that state, the lines on the road hypnotized me. I wasn't to blame. I played no part in it. Even if I did, I'm really just a child at heart, so don't you go on a-blaming me if my brain maturation hasn't caught up with my skill at making excuses.

"No, wait. Scratch that. I almost forgot. The peanuts, they was for my wife. She has these here cravings something fierce. Yes, that's it—my wife, she's pregnant. She was screaming cuz she's in labor, and I was taking her to the hospital. Could you turn on your siren, officer? Reckon we can't be late. Y'all wouldn't want to be responsible for my kid being born in this here junk heap, would you? The heck with that ticket—don't y'all have no free will?"

As any good skeptic could tell you, where there isn't a will, there's a victim and an excuse. Sadly, free excuse, like untruths in general, gets complicated—too much to remember. The arguments go on and on. Free will keeps things simple. "I was speeding, officer. Please have mercy on me. Don't give me what I deserve—spare me the karma." Every billiard ball's excuse for moving is a cue stick, a cushion, or another billiard ball. Every idiot's excuse for moving is having brains like a billiard ball—determined by the indeterminate whims of atoms.

Anonymous Authorship

Consider a recent book, *Free Will* by Sam Harris. Sam was three years old when I began attending what would become our shared alma mater: UCLA. Forgive my informality, but I can't help but think of him as a younger sibling. Per our definition, Sam is a young idiot rather than an old fool like me. He's an accomplished biomythologist, adept at using scientific data to sell his case for hard determinism without even thinking (which is good because if his hard determinism is true, he is reacting, not thinking). He writes, "We do not know what we intend to do until the intention itself arises. To understand this is to realize that we are not the author of our thoughts and actions in the way that people generally suppose."[399] The skeptic may wonder who authored this argument. We can hardly blame Sam for wanting to maintain his anonymity.

Predestined Psychopaths

At one point, Sam relates a tale of two tenderhearted psychopaths who slaughter the better part of an innocent family—husband, wife, two daughters. About the incident, he laments:

> As sickening as I find their behavior, I have to admit that if I were to trade places with one of these men, atom for atom, I would be him: There is no extra part of me that could decide to see the world differently or to resist the impulse to victimize other people.[400]

The argument is nothing new. One wonders if Sam has been inspired by Shakespeare's Isabella pleading for her brother's stay of execution in *Measure for Measure*: "If he had been as you, and you as he, / You would have slipped like him; but he, like you, / Would not have been so stern." What Isabella lacked was the chapter, line, and verse of a good scientific footnote! That Sam could swap quarks with someone else and get all the spins just right is biomythological fantasy at its best, but Sam's empathy, though a bit one-sided, is touching. As he wrote in *Letter to a Christian Nation*, "Only then [when we quit lying to ourselves about reality] will the practice of raising our children to believe that they are Christian, Muslim, or Jewish be widely recognized as the ludicrous obscenity that it is."[401]

Slaughtering a family is only "sickening," but raising children in a faith is a "ludicrous obscenity." Luckily for the psychopaths, they murdered the family rather than converting them to Christianity, or Sam might not have been so empathetic. The skeptic may suppose that Sam—when he failed to see that, per his own argument, Christians, like psychopaths, are not responsible for their passions— had eaten a bad Dawkins or Dennett footnote for lunch and his unbalanced biochemicals had gotten the best of him.

Were it not righteously and rigorously reinforced with bogus brain mythology, Sam's argument would be tired indeed. Four hundred years ago, Hamlet, having slain Polonius and driven the courtier's daughter Ophelia to suicide, offers the first insanity plea to Ophelia's incensed brother Laertes:

> What I have done
> That might your nature, honor, and exception
> Roughly awake, I here proclaim was madness.
> Was't Hamlet wronged Laertes? Never Hamlet.
> If Hamlet from himself be ta'en away,
> And when he's not himself does wrong Laertes,
> Then Hamlet does it not, Hamlet denies it.
> Who does it then? His madness. If't be so,
> Hamlet is of the faction that is wronged;
> His madness is poor Hamlet's enemy.

Having recently discovered (as we have laughingly imagined[402]) the scientific method, Hamlet (like President Nixon) speaks of himself in the third person to increase his aura of objectivity, but he offers nary a word about the immaturity of his princely prefrontal lobes or limbic system. That Hamlet lacks Sam's background in neuroscience makes no difference. As we have seen, having pleaded free excuse and seen the error of his murderous ways, Hamlet waits a scene before he sends Laertes to join Ophelia and Polonius in hell—or wherever you end up for poisoning, suicide, or too many aphorisms.

The Death of Conscious Free Will

Sam's passion is, the skeptic may imagine, to discredit conscious free will. To keep his argument consistent he must therefore argue that consciousness is an epiphenomenon, like the buzz of a bee's wings. In other words, consciousness contributes nothing to our actions, just like the buzz contributes nothing to the bee's flight. To defend this passion-laced story, Sam knows exactly who to footnote. Everyone does. I'm referring, of course, to the increasingly famous (in biomythological circles, at least) experiment carried out by physiologist Benjamin Libet (1916–2007). The skeptic knows to look for it being footnoted wherever free excuse is sold.

Libet[403] trained experimental subjects to watch a large dial with a moving hand, make a decision to flick a wrist, flick it, and note, using the dial, when the decision had been made. The experimenters then correlated all this with measured brain activity in the motor cortex (assuming that where perception is concerned, brain-time relates to mind-time better than brain-quarters relate to mental-quarters).[404] Common sense would tell us that the conscious decision would be made before the neural impulse to flick the wrist occurred. Instead, subjects' reports suggested their decisions had been made 300 milliseconds *after* the neural impulses began—they were already into the act of moving before becoming conscious of the intention to do so. Therefore, the idiots proclaim with glee, "Consciousness is an afterthought, and conscious freewill does not exist; it's a foolish illusion, and we've got the brain science to prove it. Our idiocy is mightier than your foolishness. Case closed. Nah, nah, nah, nah, nah! Told you so!"

Habit Versus Consciousness

Before we slit our foolish throats, we need to take a look at two concepts: habit and consciousness. When not reflecting on our quiet desperation, we lead lives of largely unconscious habit. We sit in the same easy chair, sleep on the same side of the bed, write with the pen in the same hand, and—with luck—kiss the same spouse. We are creatures of habit. Consciousness crops up only when habit fails; then the love affairs begin as we mistake our love of consciousness for the novelty that created it.

Stimulus-Response Seeing: An Experiment

Without sleeping on the wrong side of the bed or, worse yet, in the wrong bed, let's perform an experiment to show what I mean about the difference between habit and consciousness. Get a piece of paper and copy the word you find in the box below. If this is inconvenient, then raise your hand and write he letters in the air.

$$\boxed{\textbf{educate}}$$

When you have completed this experiment, continue reading.

If you are like most, you looked at the word once and then, relying on habit honed years ago, wrote the word nearly as easily as you sign your name. Provided you are in the habit of following instructions, writing is an example largely of automaticity or habit. It requires no real consciousness—just perhaps a gentle nudge at the beginning of the action.

I would refer to such habits as *stimulus-response* because you received the stimulus—in this case, my request for you to copy the word in the air—and automatically made the response. If, however, you are in the habit of defying instructions, then free will may have been required to break your habit of defiance. Pavlov's dogs drooling is an example of such stimulus-response writing, if not the agreement to write. Jascha Heifetz's technique when playing the violin is another example of such activity, as is the swing of Tiger Wood's golf club.

Zombie Behavior

The biologist and engineer Christopher Koch prefers to describe such responses as zombie behaviors, because they are performed largely unconsciously (zombies, by definition, performing without consciousness):

> Neurologic and psychological sleuthing has uncovered a menagerie of specialized sensory motor process. Hitched to sensors—eyes, ears, the equilibrium organ—these servomechanisms control the eyes, neck trunk, arms, hands, fingers, legs, and feet, and subserve shaving, showering, and getting dressed in the morning; driving to work, typing on a computer keyboard, and text messaging on your phone; playing a basketball game; washing dishes in the evening; and on and on. Francis Crick and I called these unconscious mechanisms zombie agents.[405]

Elsewhere in his book, Koch explains the job of "zombie agents" in a simpler way:

> This [that we live most of our lives 'beyond the pale of consciousness'] is patently true for most of the sensory-motor actions that compose our daily routine: tying shoelaces, typing on a computer keyboard, driving a car, returning a tennis serve, running on a rocky trail, dancing a waltz. These actions run on automatic pilot, with little or no conscious introspection. Indeed, the smooth execution of such tasks requires that you not concentrate ... Whereas consciousness is needed to learn these skills, the point of training is that you don't need to think about them anymore; you trust the wisdom of your body and let it take over. In the words of Nike's ad campaign, you 'just do it.'[406]

Awareness Versus Consciousness

So what is the difference between conscious behavior and automaticity/stimulus-response/zombie behavior? Sometimes the distinction is made between being *aware* of something or *conscious* of something. To explain, let's suppose you arrive at work with your keys in one hand, your cell phone in the other, a dazed look in each eye. It turns out that, during your drive to work, your thirteen-year-old daughter called to announce that she is pregnant. The fact that you arrived at work proves that you were *aware* of the lines on the road. It is less likely that you would have been *conscious* of the lines on the road.

Consciousness is not required for awareness. There is no better way to mess up a player's tennis than to get her *conscious* of her swing. Stimulus-response awareness is faster than consciousness. Consciousness is the crown jewel of exploration and can be stimulated with a question as simple as, "What does your chair seat feel like against your rear end?" Such a question, if you took a moment to answer it, would require you to add consciousness to all the other stimulus-response neurological functions required for you to sit on, rather than falling off of, the chair.

Another example of consciousness would be to turn your phone upside-down—display locked in position—and text. At first, the need to explore would demand your consciousness. In time, you would learn a new stimulus-response performance and text with wild abandon, your keypad upside-down.

David Cook

Conscious Seeing: An Experiment

To see the difference between habit and consciousness, repeat our *educate* experiment, but this time copy the word with each letter upside-down and backward, as is demonstrated in the box below.

educate

After completing the experiment, answer some questions: How many times did you have to look at the word compared to when you copied it in your habitual-stimulus-response-zombie fashion? How, exactly, did the task feel compared to the first experiment?

Practice unmakes consciousness. It takes but a moment of practice to transform conscious behavior back into zombie behavior. If you, for instance, practice writing the first *e* in *educate* upside-down and backward a half-dozen times or so, then the copying will become as unconscious as it is effortless. This is the same process you went through when you learned to write in kindergarten. It's the same process you used when you first learned to sit up or crawl or walk or drive. It's the same process you go through each time you add a word, concept, or belief to your life; practice rapidly dissolves consciousness into habit.

Hume, Novelty, and Consciousness

Relating this to free will, in *A Treatise of Human Nature*, Hume provides a useable definition of the will: "[T]he internal impression we feel and are conscious of, when we knowingly give

342

rise to any new motion of our body, or new perception of our mind."[407] If we define *new* to mean "not used or not previously used" or "used for the first time," then this definition fits nicely with what we have just been discussing. A movie is new only the first time we see it; the theory is new only the first time we grasp it; a road is new only the first time we travel it. The road previously untraveled captures consciousness; the road less traveled, less consciousness. Think, for instance, of how long it seems to take the first time we go to a new destination compared to how quickly the same trip passes when traveled daily. *New* is practically synonymous with *conscious*. When the ritual becomes habitual, we are hard-pressed to squeeze any more consciousness from the procedure. Free will is called into play most often when we contemplate defying, not following, habit. Sleeping on the wrong side of the bed demands free will. Habitually sleeping in the wrong bed may demand no free will at all. Breaking the habit, however, may require as much free will as giving up smoking, murder, gossiping, or writing polemics.

Why Libet Did *Not* Study Conscious Free Will

Returning to Libet's experiment, the task he demanded of his subjects was complex. They had to learn to coordinate their visual perception with wrist-flicking and monitored thought. It is likely they thoroughly practiced the procedure before the experiment started. During the experiment itself, many trials were averaged. Libet studied zombie performance, not consciousness.

Sure, a professional tennis player's consciousness lags behind her performance. So tell us something we don't already know about stimulus-response behavior. That player used consciousness to correct errors back in those thousands of hours when she was programming her swing, not during a professional match. The same is true of focused meditation. It's something you learn to do, until you can step outside of the noise and pictures of the mind long enough to be conscious for real, anytime you choose to flick the switch. Conscious meditation is not something supplied by your genome or operant conditioning from without. It's something you practice and learn from within.

Understanding the difference between conscious and habitual thought, zombie performance and stimulus-response, is critical in understanding free will. It matters little that we can cite that our sun is at the center of the universe or that our neurons are at the center of our thinking. Today's neuroastrology could be as far off as Copernicus's infantile astronomy. Reading our fates in the neurons or the stars, it's much the same. To provide an updated paraphrase of the words Shakespeare lent to his lean and hungry villain Cassius:

> Men at sometime were masters of their fates.
> The fault, dear Brutus, is not in our neurons,
> But ourselves, that we are underlings.

Caesar, of course, saw through the villain, observing, "He thinks too much. Such men are dangerous." Fools agree.

Foolish Agents

Consider some fools' arguments. In "Human Freedom and the Self," Roderick Chisholm defines the term *agent*:

> We should say that at least one of the events that are involved in the act is caused, not by any other events, but by something else instead. And this something else can only be the agent—the man.[408]

At our house, the events are more often caused by the woman, but besides this philosophical oversight, I believe Chisholm has a good point, one that leads, with a little philosophical sweat, to another point, that "each of us, when we act, is a prime mover unmoved. In doing what we do, we cause certain events to happen, and nothing—or no one—causes us to cause those events to happen."[409]

So who may this man or woman be, this entity capable of programming brains one tiny step at a time so we may sit up, crawl, walk, run, say words, use the words in sentences, transform the sentences into beliefs, meditate or equivocate—using steps as tiny as those of Darwin's gradualism to convert our actions into habit? Put another way, if I were to have a stroke and lose my memories, beliefs, and words, I would still be me, the agent able to put my Humpty-Dumpty's mind back together again by relearning how to react with my newly altered brain in my nearly lost world. Still, who is this phantom agent capable of programming or reprogramming brains?

345

Foolishness Emerges

In 1981, Richard Sperry gave his 1981 Nobel lecture, "Some Effects of Disconnecting the Cerebral Hemispheres":[410]

> "The events of inner experience, as emergent properties of brain processes, become themselves explanatory causal constructs in their own right ... The whole world of inner experience (the world of the humanities) long rejected by 20th century scientific materialism, thus becomes recognized and included within the domain of science ... Basic revision in concepts of causality are involved in which the whole, besides being 'different from and greater than the sum of the parts,' also causally determines the fate of the parts.

Sperry took nature badly out of context. His experiment included altered brains, altered eye movements, and altered environments—any one of which being capable of producing illusion. Despite the assumptions of his science, his metaphor outclasses the 3.5 pounds of neuro-self popular with biomythologists. As Sperry relates, an "emergent" is "different from and greater than the sum of its parts." In other words, consciousness emerges from the brain much as water emerges from hydrogen and oxygen. But who, except for after the fact, would have predicted that two gases could make a liquid?

Breaking Free Excuse

Combining the myths of Chisholm and Sperry, we find that the emergent, the agent, and the prime mover unmoved could easily be one and the same thing. If the emergent is more than the sum of its parts, it transcends math, logic, the scientific method, and—most of all—free excuse. Consciousness thus transcends the two plus two of determinism like Niagara Falls rises above the two hydrogens plus one oxygen of the periodic table.

How can the determinism of physics be reconciled with the free will of the individual? It can't. Human consciousness is not rocket science. In keeping with Kant's words, "Freedom in the practical meaning of the term is the independence of our power of choice from coercion by impulses of sensibility."[411] The agent is not the sum of its brain impulses. The agent is free to control those impulses, to drum those modules, to tell its stories, to hear or not hear its muse, to see or not see or manipulate or not manipulate its mental images, to set or not set rolling its previously constructed paths of action, thought, and heredity.

The emergent provides a natural explanation of how we as human beings transcend the chain of determinism as surely as quantum mechanics on the other end of the philosophical food chain. You called upon that free will when first you tried to write *educate* with the letters upside down and backward. Free will is not something you have; it's something you learn. The agent has free will in determining the habits of the brain one tiny, often frustrating step at a time, no step allowed to be skipped, no step possible without building on those molded by free will in the past. The skeptic suspects that you have a brain. What you do with it, how exactly you build it in the pursuit of new habits, is a matter of free will. Your ideal karma was built by your use of free will in the past. To that extent, your freedom to act depends on how you programmed your brain in the past. And if you are certain of that, then perhaps you need to reread this primer on skepticism.

In summary, I have picked on Sam Harris needlessly, and I apologize. As the skeptic may surmise, maybe Sam and I are both right. Maybe I was predestined for free will, he for free excuse. Maybe idiots were determined to be determined; fools were graced with the freedom to learn to silence their minds so as not to be ensnarled by the seductive whisperings and tantalizing images of inner muses. Maybe if Sam and I were made up of the same quarks with the same spins, then we would be offering the same spin and have the same quirks as well. We'd both be fools, or we'd both be idiots. We'd both have free will or free excuse. Our matter would demand it. In biomythology, matter is all that really matters. We can be ideally free only until we need an excuse. Then, the arguments of biomythologists can be priceless, if not exactly free—as any good skeptic suspects.

19

The Golden Rule of Skepticism

The Golden Rule

Not even the skeptic can doubt that the Golden Rule is a brushstroke swashing across religions from East to West. Confucius wrote, "What you do not want done to you, do not do to others." The Jewish sage Hillel interpreted, "That which is hateful to you, do not do to your fellow. That is the whole Torah; the rest is the explanation; go and learn." The Buddha insisted, "Do not hurt others with that which hurts you." Mohammad allowed, "None of you is a believer until you love for your neighbor what you love for yourself." In the Hindu epic, the Mahabharata, we find, "This is the sum of all duty: do nothing to others which if it were done to you would cause you pain." Jesus, in the Sermon on the Mount, says, "Therefore all things whatsoever ye would that men should do to you, do ye even so to them: this is the law of the prophets."[412] While Mohammad's and Jesus's positive rather than negative spin on the rule adds sins of omission to sins of commission, curtailing our actions *and* inactions,[413] the Golden Rule is pervasive; it respects the lines between cultures no more than clashes between cultures respect the Golden Rule.

Fortunately, history has become the Golden Rule's most valiant champion. Unfortunately, as the skeptic may suspect, history is often a masochist with a death wish. Inviting its own torture and murder at the hands of ruthless historians, history does unto others, murdering and torturing with a vengeance. Apparently inspired by history rather than Jesus, Christians abandoned the Golden Rule for crusades, inquisitions, and, during the Reformation, a hundred-year spree of human slaughter to prove who got God's story right—the one about loving neighbor and enemy.

The Enlightenment

In time, the less bloodthirsty imagined they could do better. They began an intellectual movement championing progress rather than tradition, freedom and equality rather than monarchy, and the scientific study of nature rather than the authority of religion. The Enlightenment, as this revolution in storytelling came to be known, allowed rationalism to triumph over such mysticism as the Golden Rule and loving neighbor as self.

As skeptics might have suspected, history couldn't handle reason any better than it had handled the Golden Rule. However valid history's logic, however infallible, its basic premise remained the same: mortality. All must die. Soon reason was misspelling *progress* as T-E-R-R-O-R. The maw of the French Revolution devoured its own children as the Enlightenment's first graduating class received their diplomas on the scaffold of the guillotine. Still, the Enlightenment was young, having scarcely reached the age of puberty, much less the Age of Reason. It was hoped that it would outgrow its immaturity. Maturation was slow. The enlightened, like youth in general, continued to confuse the speed of intelligence with the velocity of wisdom. They used their considerable science-engendered horsepower to race in the wrong direction—toward progress, not love and kindness.

When it comes to values, all directions have their pitfalls. In this world, the most dangerous destination of all is utopia. Only God, untrammeled by common sense, knew better than to try to cram heaven into planet Earth. The lesser gods of reason shared no such compunction.

Reason grew. In the twentieth century, imperfection was all that stood in the way of rationality's perfect secular utopia. As luck would have it, Darwin, in his "Variation under Domestication" chapter, had already proposed a solution: "Seed raisers ... go over the seed beds and pull up the 'rogues,' as they call the plants that deviate from the proper standard." Darwin banned unnatural acts from his theory but not his followers, and so the real cultivation began, this time in the name of progress rather than the name of God. According to R. J. Rummel in his study *Death by Government*,[414] in addition to the over thirty million deaths by war in the twentieth century, governments murdered 169.198 million people.

Marxism was initially promoted as economic science, at least until Stalin and Mao ran experiments falsifying the theory.

Stalin's adherents murdered 42.672 million human beings; Mao's, 37.828 million. Whether or not Rummel's figures are significant to five digits, they were certainly significant to the families and friends of the 80 million who died. Another falsification, this time of reason itself, was provided by Adolf Hitler. According to Rummel, Hitler's government murdered 20.946 million human beings, 5.291 million of whom were Jews. Some contest Rummel's figures. Being no historian, I will leave the particulars to biographer A. N. Wilson, who discusses "Hitler's crude belief in science" and "unhesitating belief in modernity":

> Hitler's zest for the modern, his belief that humanity would become more reasonable when it had cast off the shackles of the past—olde-tyme handwriting, religion, and so forth—and embraced science and modern roads, was a belief shared with almost all forward-thinking people at the time, and it continues to be the underlying belief-system of the liberal intelligentsia who control the West. His belief led directly to genocide and devastating war. [415]

Wilson continues to argue how Hitler misused reason as others have misused religion to demonstrate "with the most terrifying skill that humanity can be seduced without much difficulty into acts of collective insanity":

> While the commonplace, ordinary side of Hitler insisted that the human race had come of age, that it was now led by reason, not mumbo-jumbo, that it was rational and scientific, the extraordinary Hitler, the Mage-Hitler, the Wizard Hitler, demonstrated the exact opposite to be the case. His career showed

that human beings in crowds behave as irrationally in modern times as they did in the Dark Ages.[416]

Nietzsche feared that modernity had killed God; Wilson fears that Hitler killed modernity. Is it superstition to believe that humans are capable of combining logic and empiricism any better than they are capable of following the Golden Rule? Have we learned our lesson? Is the promise of a secular utopia once more within our reach? Between prenatal testing, genetic engineering, and political correctness, does imperfection stand a chance?

The Value of Science Versus the Science of Values

Despite history's botching of the Golden Rule and reason, the biomythologists persist in believing that, armed with the fMRI, they can do a better job in finding the answers to the difficult questions. Take fledgling neuroscientist Sam Harris plunging from the nest:

> I will argue, however, that questions about values—about meaning, morality, and life's larger purpose—are really questions about the well-being of conscious creatures. Value, therefore, translate into facts that can be scientifically understood ... The most important of these facts are bound to transcend culture ... And if there are important cultural differences in how people flourish ... these differences are also facts that must depend upon the organization of the human brain. In principle, therefore, we can account for the ways in which culture defines us within the context of neuroscience and psychology.[417]

Examining this introduction to the infant field of neuro-morality, we wonder if Sam trusts his technically adept auto mechanic to select the destinations and itineraries for the Harris vacations. Sure, the mechanic may offer advice on speeds, mileage, tank size, and tire rotation, but can he predict if Sam will get the best values for his moral landscaping at the Grand Canyon, Niagara Falls, or the Las Vegas Strip? The skeptic should wonder.

I am not alone in my skepticism about neuroscience having the potential to revolutionize society outside of providing cures for closed head injury and Alzheimer's. Oxford University philosophy professor Daniel Robertson puts it this way when discussing the identification of the mind with the brain:

> Suppose it is the case that there is nothing in the domain we are pleased to call the mental that is anything but a collection ... of physical events within the cranium ... The emphasis here is on nothing. In other words, everything that is thinkable, everything that is felt or desired, everything that is known— all this and more turns out to be no more than expressions of physicality. Fine! In that case, nothing whatever changes. Thought, desire, feeling, social interactions, and civic life—all of the problems and possibilities of life as it is actually lived—remain exactly what they are; for by the very terms of the alleged identity, they must be so.[418]

Indeed, how will learning the siren songs of the neurons change our values and reorganize our societies for the better? The skeptic might wonder if we are not mixing our metaphors, confusing our stories, promoting our false analogies. As I wrote in my introduction, "Universal explanations too often dissolve into myth when transported from one universe to another—myth and pain." Social Darwinists, such as Hitler, transported a story meant for biology into a story applied to society. The result was, indeed, myth and pain.

Transporting a story meant for the repair of brains into a story meant for the repair of societies could be no less devastating. The leap of metaphor could provide a license to replace wafers with psychoactive drugs; an excuse to chemically alter anyone whose "political correctness brain module" is acting up; an excuse to substitute hemisphere, lobe, and synapse for book, chapter, and verse; a *scientific* reason to replace the sacred stories of your choice with the latest election results of the *DSM*. Will we have to genuflect as we recite the creed of the scientific method? The skeptic might imagine a revised great chain of missing links with science on the top and us lab animals below.

But, enough of my straw-man neuroscience! Let's turn to the more modest views of *New York Times* bestseller and journalist Chris Hedges. In his book, *I Don't Believe in Atheists*, Hedges softly cautions against those who use the name of science for their ideological agendas:

> The cult of science promises to eradicate or reform the tainted and morally inferior populations of the human race ... The cult always becomes absurd. This is why, in the name of science, the regimes in Nazi Germany and the Soviet Union became scientific backwaters ... The cult of science foists upon science an impossible task—that of transforming human

nature. When science fails to achieve this goal, as it always will, science is discarded and replaced by gimmicks dressed up as science.[419]

I won't suggest that Hedges's "cult of science" sounds much like my biomythology. Whether there is any resemblance, the skeptic must decide.

In fairness to science, Sam Harris' magical rhetoric is probably not the best example of science. Polymath Stephen Jay Gould separated science from religion. He saw them as occupying "non-overlapping magisteria," separate domains of professional expertise:

> The net of science covers the empirical realm: what is the universe made of (fact) and why does it work this way (theory). The net of religion extends over questions of moral meaning and value. These two magisteria do not overlap, nor do they encompass all inquiry (consider for starters, the magisterium of art and the meaning of beauty). To cite the usual clichés, we get the age of rocks, and religion retains the rock of ages; we study how the heavens go, and they determine how to go to heaven.[420]

In his essay "The Most Unkindest Cut of All," Gould elaborated on science's role in determining values:

> Science can supply information as input to a moral decision, but the ethical realm of 'oughts' cannot be logically specified by the factual 'is' of the natural world—the only aspect of reality that science can adjudicate. As a scientist, I can refute the stated genetic rationale for Nazi evil and nonsense. But when I stand against Nazi policy, I must do so as

everyman—as a human being. For I win my right to engage in moral issues by my membership in *Homo sapiens*—a right vested in absolutely every human being who has ever graced this earth, and a responsibility for all who are able.[421]

Real scientist, biomythologist, philosopher, theologian, novelist, scholar, artist, historian—each, when deciding the fate of our culture, brings different ideas and perspectives to share, but each comes as Everyman, armed not just by expertise but by the transcendence of being human. In that best of all possible worlds, we would leave the word *scientific* outside the debates. Each storyteller would be called upon to contribute whenever culture is discussed. Such a balance of power, the skeptic might or might not accept, is to be preferred. I base this in part on the following quote by R. J. Rummel sharing his skeptical conclusion about the 169,198,000 murdered by government in the twentieth century:

> Power kills; absolute Power kills absolutely. The new Power Principle is the message emerging from my previous work on the causes of war ... and from this book on genocide and government mass murder—what I call democide—in this century. The more power a government has, the more it can act arbitrarily according to the whims and desires of the elite, and the more it will make war on others and murder its foreign and domestic subjects. The more constrained the power of governments, the more power is diffused, checked and balanced, the less it will aggress on others and commit democide. At the extremes of Power ... totalitarian communist governments slaughter their people by the tens of millions; in contrast, many democracies can barely bring themselves to execute serial murderers.[422]

The skeptic might imagine that a diversity of viewpoint, creating uncertainty and doubt and halting "progress" when necessary, is our best vigilance against the problem of evil. True, skepticism can kill progress. Still, skepticism without a doubt becomes a necessity when the story of progress kills. One only wishes that in the twentieth century the brakes of skepticism had been applied to the passionate certainty of the Enlightenment's runaway train of progress on the track bound to Utopia.

So long as we do not follow Othello's example, skepticism allows all stories while delaying the passionate actions that cannot be undone. The stories of biomythology are stories like any other. When they inspire, it should be because they are good stories, not because they are *scientific*. At least since the time of Homer, we have been inspired by stories. The good skeptic might conclude that, until all the stories are in, it is best not to place all the eggs of our culture in one basket—certainly not the basket of biomythology, and certainly not a basket beyond the reach of skepticism.

To my mind, the Golden Rule of skepticism is to never let the beauty of our stories justify the brutality of our actions. In other words, it is one thing to die for our stories, as did Socrates, Jesus, Gandhi, and King. It is quite another to kill for them, as did Alexander, Caesar, Napoleon, and Hitler. For Wittgenstein, "Philosophy is a struggle against the bewitchment of our understanding by the resources of our language."[423] For a skeptic of natural philosophy, doubt should be a battle against the bewitchment of our humanity by means of rhetoric, with or without the modifier *scientific*.

20

Skeptical Faith

Like Hume, Peter Unger is a bona fide radical skeptic—a real one, not some namby-pamby turf skeptic whose doubt somehow bypasses the potential omnipotence of the tools of science. When it comes to rescuing skepticism from the abyss of certainty, Unger has few rivals. He argues that to *know* something in the fullest sense of the word we have to be *certain* of it in the fullest sense of the world. If someone were to say to you, "I know it is raining out, but I'm not so certain that it is raining out," you would be correct to doubt the extent of her knowledge. If you are not *certain* it is raining out, you might have faith enough in probabilities to carry an umbrella, but you certainly would not *know* it is raining out any more than you know it's evolutioning cats and dogs—or so Unger might argue.

How certain must certainty be? For Unger, *certainty* is an absolute term, *absolute certainty* being but a redundancy. Anything I am less certain of than I am certain of habitual action in a habitual environment, I am not certain of at all. If I watch carefully as I press the tips of my pointer fingers together until I am certain they are touching, then anything I am less certain of no longer qualifies as certainty. Sure, what we call atoms have the power to propel submarines and vaporize women and children, but can I be

360

equally certain that Rutherford got the uncuttable atom right? Can I be equally certain that Bohr got the uncuttable atom right? Can I be equally certain that the atomic zookeepers have no cuts left to make or that the uncuttable atom is not merely a mathematical Halloween costume of ultimate reality masquerading for the limits of the human mind?

And how about Lamarck's or Darwin's or Dawkins's or Gould's version of evolution—can all be true, like Christianity, Buddhism, light waves, and light particles all claiming to be the light of truth?[424] Can I be as certain of each version of evolution as I am certain that my pointer fingers touch when I press them together? The increasing depth and number of my wrinkles seen in the mirror assures me that my face, at least, has evolved over my six and a half decades of time. While I am certain that the survival of the fittest can always be used to concoct a tale about why the fittest wrinkle survived, I have no certainty of why life in general seems to have changed over the eons. Unless I am as certain that in the last billion years no genetic material has been spontaneously generated by the mysteries of the quantum foam, no DNA has made it to earth from space—sent by Martians in silicone pain or sent from the inner sanctums of marauding comets—as I am certain that my pointer fingers are touching the keyboard before me, then I do not *know* that our microbe-to-marching-bands theory is correct.

Without faith, how certain are any of us about the things that we claim to know? Peter Unger might ask us to look at a

David Cook

tabletop. Is it flat? Are we certain? Again, for Unger, flat is an absolute term. Tables are either flat or they are not. If I were to pull out my trusty titanium straightedge and trusty electron microscope, are we certain I would not find the tabletop is irregular? So if we are not even certain that the tabletop is flat, how much can we say with certainty how much we certainly know?

Unger uses such an example of "flat" to highlight the absolute character of "certain":

> As a matter of logical necessity, if someone is certain of something, then there never is anything of which he is more certain. For ... if the person is more certain of any other thing, then either he is certain of the other thing while not being certain of the first, or else he is more nearly certain of the other thing than he is of the first; that is, he is certain of neither. Thus, if it is logically possible that there be something of which a person might be more certain than he now is of a given thing, then he is not really certain of that given thing.[425]

If such certainty is necessary for knowledge, how much can we really claim to know? In Unger's words, "That, in the case of every human being, there is hardly anything, if anything, at all, which the person knows to be so."

English philosopher, Herbert Spencer (1820-1903), who coined "the survival of the fittest," also defined science as "organized knowledge." [426] If Unger and Socrates are correct, there is very little to organize. As you recall, the scientific logic known as induction cannot provide *certain* knowledge until all the information is in,

362

freeing induction from the risks of future discovery. If induction has failed in the past and induction cannot be trusted to fail in the future, then the past does not predict the future and induction cannot be trusted. *Uncertain knowledge* is, at best, an oxymoron. Somewhere a black swan or married bachelor could be lurking in the imperfect fossil record of the Precambrian. By definition, so long as the world is not closed to discovery, scientific knowledge is open to uncertainty, or it wouldn't be science. Therefore, what do any of us (biomythologists included) really know?

Malevolent Demons and God

As the Enlightenment began to percolate, the man largely responsible for founding science on the bedrock of doubt was mathematician and philosopher René Descartes. Not only did Descartes show us a path between geometry and algebra, but also he fathered science in France even as Sir Francis Bacon was stealing Shakespeare's priority in England. Descartes decided to start the whole scientific and philosophic endeavor from scratch. He began by asking himself what he could really accept without doubt. He came up with the famous "I think, therefore I am." His thoughts could all be delusions, but delusion or not, he was experiencing the thoughts, so he must exist—for without existence, without being, he could experience nothing.

How did Descartes get around the possibility of his thoughts all being delusions or dreams created by "some malevolent demon"?

How did this father of rational scientific thought know he was not being deceived in perceptions confirmed by careful comparison of all his senses? How did he know that there are not only stories, but actions as well? Descartes explained, "For since God is no deceiver, it necessarily follows that I am not herein deceived."[427]

For Descartes, certainty in perception followed certainty in God—otherwise there was no argument against the ultimate ravages of skepticism. Plato similarly knew that *Truth* had to be capitalized; he bet his drachmas on transcendent Forms, not the paltry geometry forms tarnishing here on planet Earth. As I have suggested in previous chapters, perception, induction, language, analogy—all may be local phenomena, crumbling when extended to new perspectives. Until all the data is in, we may be like dessert nomads constructing our universals out of sand and camels. Until then, chaos control must depend on *something more*.

Wittgenstein recognized that it was "illusion that the so-called laws of nature are the explanations of natural phenomenon."[428] Sure, we know what energy *does*, but we cannot really explain, as Richard Feynman noted, what energy *is*. Just as the uncuttable atom can be cut and cut and cut, everything regresses to mystery; call it what you will. Everything progresses to faith in that mystery. Without postulating an absolute, can the persistent whys of philosophers and four-year-olds never be silenced? From the day Nietzsche reported God's death, without making the wounds visible on the Internet, the hunting season on biomythologists has been reopened to any universal skeptic with a doubt.

Running in Skeptical Circles

If Unger is correct and certainty is an absolute, then I cannot be certain that the table, which I can examine with my own fingertips, is flat. How then can I be certain that any premise that eludes the reach of all fingertips is true? What can I really *know* about logic? Bragging about past successes of logic in bringing us the modern world is support for pragmatism, not logic. Moreover, if skepticism is no more than logic in which all premises are uncertain, then how can I be sure of skepticism? Can I be as certain of logic and skepticism as I am certain of the nose on my face? As our example of Darwin's tutu demonstrated, validity is not Truth. Can I be as certain that this time validity will suffice as I am certain that I am in pain when I stub my toe? Can I have certainty that the words can precisely capture premise when the uncuttable atom has been diced into a zoo of particles, charity has been stretched into every conceivable mode of torture, and holy wars have been excused as tough love? Can I be any more certain of skepticism?

We can argue in circles, using reason to defend reason from doubt, but if skepticism is valid shouldn't we be skeptical of skepticism? If humans were ninety-nine percent ape, why would we trust their theories? Could Freud's *unconscious mind* be merely an artifact of Freud's unconscious mind? If determinism is true, then is not the determinist's reasoning but a continuation of a neural chain reaction? How can we dismiss metaphysics without using metaphysics? Where is the scientific study (spare me the anecdotes) that proves the scientific method works as well for the kiss as the billiard ball? If analogies were not false from some perspective, would they not be identities? Have analogies become the new myths? Is using analogy to dismiss analogy, logic to dismiss logic, or reason to dismiss reason about as consistent as being intolerant of intolerance, or dogmatic against dogma? Can we universally dismiss universals? Most importantly, if language can't be trusted, then why don't I shut up?

Skeptical Faith

Just as mysticism, with its awareness of God in everything, is only one step away from atheism, so is skepticism but one step away from *fideism*, the doctrine that knowledge depends on faith or revelation, not reason. To repeat Unger's words, "That, in the case of every human being, there is hardly anything, if anything, at all, which the person knows to be so." We will remain innocent of knowledge, so long as doubt is allowed to have its reasons, which is why we are lucky to have faith. Nothing demands faith quite like doubt. Theologian Søren Kierkegaard (1813-1855), best known for coining "leap of faith," cautioned against confusing historical knowledge with faith and insisted that "faith had uncertainty as a useful teacher." Why? If you are a skeptic, if you have doubt in certain knowledge, you need faith more than ever to banish chaos.

The skeptic cannot *know* if the jet will make it, but if that skeptic would get from Chicago to Hong Kong in fifteen hours, she must take the leap of faith needed to board the jet, no matter how much reason she dedicates to comparing the airfares and safety records of the airlines. Dawkins, Shermer, and Coyne admit as much when they admit that the stories of science are provisional, not certain, truth. Thus, even in science, if we are to act on our incomprehensible brute facts at all, we must act on faith until our data is complete, for in the skeptic's story, incomplete data is too uncertain to qualify as knowledge. True enough requires only justification enough, but to know absolute truth demands absolute proof. Otherwise, all that shields us from doubt is faith.

For the skeptic, faith is the passion to find the image of truth reflected in an epistemic mirror shattered by uncertainty. I am therefore skeptical enough to classify as *faith* the passionate beliefs that most often hold together the ideal space of our stories—faith in materialism or faith in transcendence; faith in something less or faith in something more; faith in reason or faith in faith; faith in facts or faith in the human mind's ability to out-discover facts; faith in universal, necessary, and certain knowledge or faith in Sophistry, rhetoric, and different truths for Athens and Sparta; faith that the laws of nature can or cannot be suspended long enough to allow for miracles; faith that science may one day control or not control what matters even as it now controls matter; faith that the scientific method can or cannot find unchanging truth in an ever-changing world; faith that we will rejoin our loved ones in eternity or lose them to oblivion; faith that the World Trade Center floors we walk in life will still hold our weight tomorrow.

As the good skeptic believes, reason—itself the slave of passion—too often mistakes faith for knowledge, but in the world falling beyond the reach of certainty, all claims to knowledge are but pretenses embellished by the rhetoric of objectivity. Perhaps faith is as close as we will come to knowing the story of the absolute until we meet the absolute face-to-face. So skeptics remain wary. Why allow ourselves to be seduced by adjectives, even adjectives as *scientific* as waves of justified rock-hard data building into mountains before they break into dust? Despite the growing fashionableness of biomythology, a life unexamined by the fMRI may still be worth living.

Our private stories are our own. To align them with the promise of everlasting life is to escape losing those we love for as long as we possess the passion to tell stories. To align our private stories with oblivion is to live for an escape from consciousness, have a way out, just so long as Death maintains its right to be proud. Only skeptics are free to choose their form of escapism.

Speaking pragmatically, when it comes to our private stories, why shouldn't we be skeptical of the *ersatz* self of biomythology when we can enjoy the sweet-cream purpose and perpetuity of soul? We stand between the boundless mirrors of birth and death and glimpse immortality in the multitude of our reflections. Our ability to tell ourselves such stories is a divine gift. Why would we return the gift unopened? Why would we reason the gift away? Why would we sacrifice it to placate those who worship the fleeting facts and ephemeral observations of public action? The private stories we use to cushion ourselves from chaos need not rest in the actions of evidence. The proof of the story rests in the perceived softness of the cushion.

To quote Socrates shortly before he died for his stories:

> Philosophy then persuades the soul to withdraw from the senses in so far as it is not compelled to use them and bids the soul to gather itself together by itself, to trust only itself and whatever reality, existing by itself, the soul by itself understands, and not to consider as true whatever it examines by other means, for this is different in different circumstances and is sensible and visible, whereas what the soul itself sees is intelligible and invisible.[429]

The philosophers once knew. Our senses are necessary for action. When it comes to transcendence insulating us from chaos, however, we need not depend on our eyes; empiricism need play little role. It's the inner, not outer, light that allows us to see. The feel of our private stories about blue sky, true love, and eternity exceed the domain of peer review.

The skeptic should not confuse the material with the ideal, actions with stories. If Scarlett O'Hara falls down a staircase and no one sees the movie, did she move? Of course not. She did not move even if the movie theater was full. The actors or camera may have once moved, but the images from the projector do not. Individual stationary images merely appear in succession in different positions on the screen. Whether or not the illusion of motion and depth represents reality, the motion and depth themselves are ideal—stories told with image and word. The silver screen's motion and depth have no existence outside those private stories we call *mind,* and yet the 3-D embrace inspired by the flat, unmoving pictures of Rhett and Scarlett on the silver screen move us still.

When it comes to the ideal, whether or not the story conforms to the physical world, our feel of Scarlett's tumble down the stairs is true. The story of Rhett and Scarlett and Melanie and Ashley and Bonnie and Mammy, falling outside history, does not necessarily *describe* love; it *evokes* love. The best stories do, I imagine, but I could be wrong. I could be wrong about a lot of things. Most things. Everything. If, for real skeptics, all doubt is reasonable, then you are correct to have your doubts. If I have increased your doubt about biomythology, this book, or both, then I have done my job.

Are we the only creatures who can tell stories about the planet Pluto? Can we reduce all to passions, actions, and stories? Are actions material and stories ideal? Is belief but a story held aloft by passion? Where does the soul of the storyteller end and the soul of the story begin? Are the two souls one? Do they emerge from starstuff or something more? Is a story no truer than the hope and actions it inspires? Is the truest story the one that opens hearts, not closes minds? The one that inspires love, not just rockets to the moon? Is eternal life but the story that love will never end? Is a skeptical story necessary whenever stories are being told to justify necessary evils and make Death unnecessarily proud? When it comes to telling public stories, whether to theater audiences or execution squads, should we be skeptical about shouting "Fire"?

Stories are like balloons. When inflated with certain passion, they ascend as belief; when inflated with certain knowledge, they ascend as truth. The skeptic suspects that, beyond love, certain belief is cheap; certain truth, often more than we can afford. In Aristotle's words, "All men desire to know." In Socrates' words, "All I know is that I know nothing." How should the skeptic solve the dilemma that our words revolve around the desire to know and the wisdom to know better?

Burn this book.[430]

Final Assessment

Stories are the glue holding culture together. It's small wonder that everyone wants to tell the best story. One way to do this is to tell the best story. The other is to rank stories by category: abstract, abstraction, account, allegory, analogy, argument, biomythology, category, chronicle, complaint, daydream, delusion, description, discovery, dissertation, dogma, drama, dream, episode, excuse, evaluation, explanation, fable, fact, fairytale, fallacy, fantasy, fiction, Gospel, history, hypothesis, idea, illusion, interpretation, lie, logic, meme, memory, metaphor, myth, narration, news, nursery rhyme, parable, paradigm, philosophy, pitch, picture, plea, polemic, prayer, prediction, propaganda, prophesy, purpose, reality, reason, revelation, rhetoric, romance, science, script, scripture, synopsis, tale, thought, truth, or universal—to name a few.

Some hope that the category can rescue a story from skepticism. The skeptic sees such hope as vanity. A category— being itself but a story, a visitor from the ideal, a muse's best guess—is equally open to skepticism. Great stories change the world, not because they can be categorized as science or scripture, but because they are great stories, inspiring us to overcome our fear of gravity and empire long enough to feel the transcendence of what it means to be human. The less-than-great stories in these pages must stand on their own merits, rather than on category. Indeed, they traverse most of the categories listed above, yet their success rests on their ability to inspire skepticism: to do for Darwin and science what Richard Dawkins' story has attempted to do for God and religion.

Have we rescued skepticism from toppling into the abyss of certainty, balancing doubt about stories of the natural and supernatural? Have we encouraged skeptics to step into the future and look back to doubt today's starstuff-fossil-genome-and-neuron stories just as today we look back at the air-earth-fire-and-water stories of yesterday? Such doubt is, of course, necessary only when stories are used to bar the way to future discoveries or justify actions that make "the life of man solitary, poor, nasty, brutish, and short." Love, joy, peace, patience, kindness, goodness, faithfulness, gentleness, self-control—against such things there is no reason for skepticism, except when rhetorical reason redefines such things as excuses for killing.

"Your *Skeptic's Guide* has failed!" some will insist, and they will doubtlessly have their stories to show why skepticism about some stories is more equal than skepticism about others. And, if their stories are sufficiently well told, we may believe them—for a time. There is only one defense against the universal solvent of rhetorical reason washing away all arguments and observations conflicting with passion. Have doubt. With skepticism, all stories remain possible.

What was the probability that the sun could cross the sky without moving? What was the probability that time and space were just matters of perspective? What was the probability that clouds of dust would transform into galaxies or that the immutable heavens were expanding? What was the probability that that rock-solid continents were drifting or that the shortest distance between two points was not necessarily a straight line? Who would have predicted these discoveries, these shifts in perspective, these plot twists in our most passionate stories of the cosmos?

Only when it comes to betting on the future, skeptics prefer to play the odds. They know that whenever the word *probability* crops up, time is deep, and time is on their side. In the long run, skepticism is the safest bet. Outside of faith in transcendence, nothing has demonstrated the staying power of skepticism. And for good reason. Skepticism is itself a faith in transcendence, a faith in future discovery, a faith in something more than our most cherished stories now reveal. The true skeptic awaits the stories of ultimate truth and glory hiding behind the darkness of the material glass.

Bibliography

Abrams, M.H., General Editor. *The Norton Anthology of English Literature.* New York: W.W. Norton and Company, 1968.

Alexander, DR and Numbers, RL, eds. *Biology and Ideology From Descartes to Dawkins.* Chicago: The University of Chicago Press, 2010.

Andrews, R. *The Columbia Dictionary of Quotations.* New York: Columbia University Press, 1993.

Aquinas, Thomas. "Question LXXXIII, Free Choice, from the Summa Theologica." In *Free Will*, by ed Derk Perboom, 34-41. Indianapolis, Indiana: Hackett Publishing Company, Inc., 1997.

Arbesman, Samuel. *The Half-Life of Facts: Why Everything We Know Has an Expiration Date.* New York: Current, a member of Penquin Group, 2012.

Aristotle. *The Complete Works of Aristotle, The Revised Oxford Translation, Volume One, Jonathan Barnes Editor.* Princeton, New Jersey: Princeton University Press, 1984.

Armstrong, Karen. *A Short History of Myth.* New York: Canongate, 2005.

Ash, Sidney. *Pretrified Forest: A Story in Stone.* Petrified Forest, Arizona: Petrified Forest Museum Association, 2005.

Bacon, Francis. *The New Organon, edited by L. Jardine and M. Silverthorne.* Cambridge: Cambridge University Press, 2000.

Bartlett, John. *Bartlett's Familiar Quotations, Fourteenth Edition.* Little, Brown and Company: Boston, 1968.

Benjamin, Libet. *Mind Time: The Temporal Factor in Consciousness.* Cambridge, Massachusetts: Harvard University Press, 2004.

Blackburn, Simon. *Oxford Dictionary of Philosophy.* New York: Oxford University Press, 2008.

Blackmore, Susan. *Consciousness: An Introduction.* Oxford: Oxford University Press, 2004.

Bowker, John. *World Religions: The Great Faiths Explored and Explained.* New York: D K Publishing Inc., 1997.

Brooks, John Langdon. *Just Before the Origin: Alfred Russel Wallace's Theory of Evolution.* New York: Columbia University Press, 1984.

Cave, Stephen. *Immortality, The Quest To Live Forever and How It Drives Civilization.* New York: Crown Publishers, 2012.

Chalmers, David J. *The Conscious Mind: In Search of a Fundamental Theory.* New York: Oxford University Press, 1997.

Charlesworth, Brian and Deborah. *Evolution: A Very Short Introduction.* New York: Oxford University Press, 2003.

Chesterton, G.K. *The Project Gutenberg eBook of Eugenics and Other Evils.* May 3, 2008.

Chomsky, Noam. *Laguage and Mind.* Cambridge: Cambridge University Press, 2006.

Comte-Sponville, André. *The Little Book of Atheist Spirituality.* New York: Viking, 2007.

Cook, David. *The Anatomy of Blindness.* Bloomington, Indiana: AuthorHouse, 2013.

Cox, Brian and Forshaw, Jeff. *Why Does E = mc²?* Cambridge, Massachusetts: Da Capo Press, 2010.

Coyne, Jerry A. *Why Evolution Is True.* New York: Viking, 2009.

Crick, Francis. *The Astonishing Hypothesis: The Scientific Search for the Soul.* New York: A Touchstone Book, 1995.

Damasio, Antonio. *Descartes' Error: Emotion, Reason, and the Human Brain.* New York: Penquin Books, 2005.

Darwin, Charles. *Darwin: The Indelible Stamp: The Evolution of an Idea. Edited, with commentary by James D. Watson.* Philadelphia: Running Press, 2005.

—. *On The Origin of Species: The Illustrated Edition, David Quammen, General Editor.* New York/London: Sterling, 2008.

Dawkins, Richard. *The God Delusion.* Boston, New York: Houghton Mifflin Company, 2006.

—. *The Greatest Show on Earth: The Evidence for Evolution*. New York: Free Press, 2009.

—. *The Magic of Reality: How We Know What's Really True*. New York: Free Press, 2012.

—. *The Selfish Gene*. Oxford: Oxford University Press, 30th Anniversary Edition, 2006.

Dennett, Daniel C. *Breaking the Spell: Religion as a Natural Phenomenon*. New York: Viking, 2006.

—. *Consciousness Explained*. New York: Back Bay Books, 1991.

—. *Darwin's Dangerous Idea: Evolution and the Meaning of Life*. New York: Simon and Shuster Paperbacks, 1995.

—. *Intuition Pumps and Other Tools for Thinking*. New York: W. W. Norton and Company, 2013.

Descartes, René. *Discourse on Method. Bol. XXXIV, Part 1. The Harvard Classic*. *New York: P. F. Collier & Son, 1909-14*. Bartleby.com, 2001.

Eccles, John C. *How the Self Controls Its Brain*. Berlin: Springer-Verlag, 1994.

Encarta World English Dictionary. New York: St Martin's Press, 1999.

Feyerabend, Paul. *Against Method*. London: Verso, 2010.

Feynman, Richard P, Robert B Leighton, and Matthew Sands. *The Feynman Lectures*. New York: Basic Books, 2010.

Feynman, Richard P. *The Pleasure of Finding Things Out*. Cambridge, Massachusetts: Perseus Books, 1999.

Fodor, J., and Piattelli-Palmarini, M. *What Darwin Got Wrong*. New York: Farrar, Straus and Giroux, 2010.

Godfrey-Smith, Peter. *Theory and Reality: An Introduction to the Philosophy of Science*. Chicago: University of Chicago, 2003.

Goldacre, Ben. *Bad Pharma: How Drug Companies Mislead Doctors and Harm Patients*. New York: Farber and Farber, inc., 2013.

Goldman, Steven L. *Science Wars: What Scientists Know and How They Know It*. Chantilly, Virginia: The Teaching Company, 2006.

David Cook

Gopnick, Adam. *Angels and Ages: A Short Book about Darwin, Lincoln, and Modern Life*. New York: Alfred A. Knopf, 2009.

Gould, Stephen Jay. *The Richness of Life: The Essential Stephen Jay Gould, Edited by Steven Rose*. New York: W. W. Norton and Company, 2007.

—. *The Structure of Evolutionary Theory*. Cambridge, Massachusetts: The Belknap Press of Harvard University Press, 2002.

Greenblatt, Stephen. *The Sweve: How the World Became Modern*. New York: W.W. Norton and Company, 2012.

Gregory, Frederick. *Natural Science in Western History*. Boston: Houghton Mifflin Company, 2008.

Grim, Patrick. *The Philosopher's Toolkit: How to Be the Most Rational Person in Any Room, Course Guidebook*. Chantilly, Virginia: The Great Courses, 2013.

Hall, D E. "Religious attendance: more cost-effective than Lipitor?" *J Am Board Fam Med*, 2006: 103-109.

Hall, James. *Practically Profound: Putting Philosophy to Work in Everyday Life*. Lanham, Maryland: Rowen and Littlefield, 2005.

Harris, Sam. *Free Will*. New York: Free Press, 2012.

—. *Letter to a Christian Nation*. New York: Alfred A. Knopf, 2006.

—. *The Moral Landscape: How Science Can Determine Human Values*. New York: Free Press, 2010.

Hazen, RM. *The Joy of Science, Course Outline*. Chantilly, Virginia: The Teaching Company, 2001.

Hodges, Chris. *I Don't Believe in Atheists*. New York: Free Press, 2008.

Hubbard, L. Ron. *Dianetics: The Modern Science of Mental Health, 50th Anniversary Edition*. Los Angeles: Bridge Publications, Inc., 2000.

Hume, David. *A Treatise of Human Nature, Edited by Ernest C Mossner*. New York: Penguin Books, 1969.

378

—. *An Enquiry Concerning Human Understanding: A Letter from a Gentleman to his Friend in Edinburgh, Edited by Eric Steinberg.* Indianapolis, Indiana: Hackett Publishing Company, 1980.

—. *Dialogues and Natural History of Religion, Edited by J.C.A. Gaskin.* New York: Oxford University Press, 1993.

Isaacson, Walter. *Einstein: His Life and Universe.* New York: Simmon and Shuster, 2007.

Israel, Paul. *Edison: A Life of Invention.* New York: John Wiley & Sons, Inc., 1998.

Jablonka, Eva, and Lamb, Marion J. *Evolution in Four Dimensions: Genetic, Epigenetic, Behavioral, and Symbolic Variation in the History of Life.* Cambridge, Massachusetts: The MIT Press, 2006.

Jardine, L., and A Stewart. *Hostage to Fortune: The Troubled Life of Francis Bacon.* London: Gollancz, 1998.

Johanson, Donald C. and Wong, Kate. *Lucy's Legacy: The Quest for Human Origins.* New York: Three Rivers Press, 2010.

Johnson, Paul. *Darwin: Portrait of a Genius.* New York: Viking, 2012.

Johnson, Phillip E. *Darwin on Trial, Second Edition.* Downers Grove, Illinois: Intervarsity Press, 1993.

Kane, Robert. *A.* n.d.

—. *A Contemporary Introduction ot Free Will.* New York: Oxford University Press, 2005.

Koch, Christof. *Consciousness: Confessions of a Romantic Reductionist.* Cambridge, Massachusetts: The MIT Press, 2012.

Kors, Alan Charles. *Voltaire and the Triumph of the Enlightenment, Course Guidebook.* Chantilly, Virginia: The Teaching Company, 2001.

Krukonis, Greg, and Tracy Barr. *Evolution for Dummies.* Hoboken, New Jersey: Wiley Publishing, Inc., 2008.

Kuhn, Thomas S. *The Structure of Scientific Revolutions.* Chicago: University of Chicago, 1996.

Lahave, Noam. *Biogenisis: Theories of Life's Origin*. Oxford: Oxford University Press, 1999.

Landesman, Charles, and Meeks, Roblin. *Philosophical Skepticism*. Malden, Massachusetts: Blackwell Publishing, 2003.

Lewis, David. *Papers in Metaphysics and Epistemology*. Cambridge: Cambridge Univeristy Press, 1999.

Lionni, Paolo. *The Leipzig Connection: The Systematic Destruction of American Education*. Sheridan, Oregon: Herrron Books, 1993.

Lisle, Jason. *Discerning Truth: Exposing Errors in Evolutionary Arguments*. Forest Green, AR: Master Books, 2010.

Malthus, Thomas. *An Essay on the Principle of Population*. London, 1798.

McGinn, Colin. *Shakespeare's Philosophy: Discovering the Meaning Behind the Plays*. New York: Harper Perennial, 2007.

—. *The Mysterious Flame: Conscious Minds in a Material World*. New York: Basic Books, 1999.

Merton, Robert K. "The Unanticipated Consequences of Purposive Social Action." *American Sociological Review*, 1936: 894-904.

Mill, John Stuartl. *On Liberty*. New York: The Liberal Arts Press, Inc. A Divsion of Bobbs-Merrill Company, Inc., 1956.

Miller, Kenneth R. *Only a Theory: Evolution and the Battle for America's Soul*. New York: Viking, 2008.

Nagasawa, Yugin. *God and Phenomenal Consciousness: A Novel Approach to Knowledge Arguments*. New York: Cambridge University Press, 2008.

Nagel, Thomas. *Mind and Cosmos: Why the Materialist Neo-Darwinian Conception of Nature is Almost Certainly False*. New York: Oxford University Press, 2012.

—. *The View From Nowhere*. New York: Oxford Univerisity Press, 1989.

Neitsche, Friedrich. *Thus Spoke Zarathustra*. New York: Barnes and Noble, 2012.

Newton, Isaac. *Four Letters from Sir Isaac Newton to Doctor Bentley Containing Some Arguments in Proof of a Deity*. London: R and J Dodsley, 1756.

Noë, Alva. *Action in Perception.* Cambridge, Massachusetts: The MIT Press, 2004.

—. *Out of Our Heads: Why You Are Not Your Brain, and Other Lessons from the Biology of Consciousness.* New York: Hill and Wang, 2009.

Partington, Angela, ed. *The Oxford Dictionary of Quotations, Revised Fourth Edition.* Oxford: Oxford University Press, 1996.

Pereboom, Derk. ed. *Free Will.* Indianapolis, Indiana: Hackett Publishing Company, Inc., 1997.

Pinker, Steven. *How the Mind Works.* New York: W.W. Norton and Company, Inc., 2009.

Plantinga, Alivin. *Where the Conflict Lies: Science, Religion, Naturalism.* Oxford: Oxford University Press, 2011.

Plantinga, Alvin. *Where the Conflict Really Lies.* New York: Oxford University Press, 2011.

Plato. *Plato: Complete Works, John M. Cooper Editor.* Indianapolis, Indiana: Hackett Publishing Company, 1997.

Popper, Karl. *The Logic of Scientific Discovery.* New York: Routledge Classics, 2002.

Powell LH, Shahabi L, Thoresen CE. "Religion and Spirituality: Linkages to Physical Health." *American Psychologist,* 2003: 36-52.

Robinson, Daniel N. *Consciousness and Its Implications.* Chantilly, Virginia: The Teaching Company, 2007.

Robinson, Marilynne. *Absence of Mind: The Dispelling of Inwardness from the Modern Myth of the Self.* New Haven: Yale University Press, 2010.

Robison, Daniel N. *The Great Ideas of Psychology.* Chantilly, Virginia: The Teaching Company, 1997.

Rosenblum, Bruce, and Kuttner, Fred. *Quantum Enigma: Physics Encounters Consciousness.* New York: Oxford University Press, 2008.

Rummel, R.J. *Death by Government.* New Brunswick: Transaction Publishers, 2008.

Ruse, Michael, and Joseph, Travis, eds. *Evolution: The First Four Billion Years.* Cambridge, Massachusetts: The Belknap Press of Harvard University Press, 2009.

Ryle, Gilbert. *The Concept of Mind.* Chicago: University of Chicago, 2002.

Sagan, Carl. *The Demon-Haunted World: Science as a Candle in the Dark.* New York: Ballantine Books Edition, 1997.

Scheiman, M., et al. "Randomized clinical trial of treatments for symptomatic convergence insufficiency in children." *Arch Ophthalmol,* 2008: 1336-1349.

Schmidt, Eric, and Jared Cohen. *The New Digital Age: Reshaping the Future of People, Nations, and Business.* New York: Alfred A. Knopf, 2013.

Searle, John R. *The Mystery of Consciousness.* New York: The New York Review of Books, 1997.

Shermer, Michael. *Skepticism 101: How to Think Like a Scientist, Course Guidebook.* Chantilly, Virginia: The Great Courses, 2013.

—. *The Believing Brain, from Ghosts and Gods to Politics and Conspiracies—How We Construct Beliefs and Reinforce Them as Truths.* New York: Times Books, Henry Holt and Company, 2011.

—. *Why Darwin Matters, The Case Against Intelligent Design.* New York: Owl Book/ Henry Holt and Company, 2006.

Silvertown, Jonathan, ed. *99% Ape: How Evolution Adds Up.* Chicago: University of Chicago Press, 2008.

Smith, Jonathan Z., ed., with The American Academy of Religion. *The HarperCollins Dictionary of Religion.* San Francisco: Harper San Francisco, 1995.

Snow, C. P. *The Two Cultures, with Introduction by Stefan Collini.* Cambridge: Cambridge University Press, Canto Edition, Thirteenth Printing, 2010.

Sober, Elliott. *Philosophy of Biology, Second Edition.* Boulder, Colorado: Westview Press, 2000.

Solomon, Robert C. *Spirituality for the Skeptic.* New York: Oxford University Press, 2007.

—. *Spirituality for the Skeptic: The Thoughtful Love of Life.* New York: Oxford University Press, 2007.

The Oxford American College Dictionary. Oxford: Oxford University Press, 2002.

The Shorter Oxford English Dictionary. Oxford: Oxford Univerity Press, 2007.

Tillich, Paul. *Dynamics of Faith.* New York: HarperOne, 2009.

Twain, Mark. *Pudd'nhead Wilson and Those Extraordinary Twins.* New York: Penguin Books, 1969.

—. *The Wit and Wisdom of Mark Twain, edited by Alex Ayers.* New York: Meridian, 1989.

Unger, Peter. *Ignorance.* Oxford: Clarendon Press, 1975.

van Wyhe, John. "Charles Darwin's Cambridge Life." *Journal of Cambridge Studies,* 2009: 2-13.

Watson, J. D., and F. H. Crick. "Molecular structure of nucleic acids; a structure for deoxyribose nucleic acid." *Nature,* April, 1953: 737-738.

Watson, James D. *The Double Helix: A Personal Account of the Discovery of the Structure of DNA.* New York: A Touchstone Book, 2001.

Williams, Thomas. *Reason and Faith in the Middle Ages, Course Guidebook.* Chantilly, Virginia: The Teaching Company, 2007.

Willis, Gary. *Head and Heart: American Christianities.* New York: The Penguin Press, 2007.

Wilson, A.N. *Hitler.* New York: Basic Books, 2012.

Wittgenstein, Ludwig. *Philosophical Investigations, The German text, with an English translation by G.E.M. Anscombe, Revised 4th edition by P.M.S. Hacker and Joachim Schulte.* Malden, MA: Blackwell Publishing Ltd., 2009.

—. *Tractatus Logico-Philosophicus.* 1918.

Woodward, Sir Arthur Smith. *The Earliest Englishman.* London: C.C. Watts & Co. Limited, 1948.

Notes

Introduction

1 Feyerabend 2010, p. 9.

2 (Jardine and Stewart 1998), pp. 460-461.

3 Excuse the wordplay. Allusions, rhymes, puns, oxymoron, aphorisms, fallacies—name your rhetorical device; I pretend no allegiance to "plain prose." This book is better suited for the literature-minded than the literal-minded. It caters to those who prefer the truth of fictions to the truth of facts. Throughout these chapters, I play with words to remind us of how words play with our minds. The *of* in *Science of Persuasion* was deliberately ambiguous, designed to seduce those who buy books for their covers. Even the definition of biomythology—"The use of the scientific method to control minds"—is wordplay. Neuroscience controls brains/minds. Propaganda borrows the mystique of science to "control minds." Thus, biomythology revolves around the uncertainty of words, positioning neuroscience—exactly where it belongs until it exceeds the sum of its correlations—with propaganda. Ultimately, this book uses rhetoric to view rhetoric, which again returns us to words for the purpose of persuasion. Finally, one rule applies throughout: never allow logic, accuracy, or common decency to interfere with a punch line.

4 Just as piano lessons train the player to coordinate ears, brain, and body to dance with the keyboard, vision therapy trains the seer to coordinate eyes, brain, and body to dance with the world. It allows us to direct action and derive meaning from light falling on the brain's only moveable parts: the retinas. Or, if you find that jargon meaningless, vision therapy is physical therapy for the eyes.

5 Feyerabend 2010, p. 14.

6 *J Am Optom Assoc, Am J Optom Physiol Opt, JOVD*, and *JBO*, to name a few.

7 (Gregory 2008), p. 269.

8 *Learning disability*: "explanation for a child who is bright in everything but school."

9 *Dyslexia*: "reading disability caused by a presumed brain dysfunction—of the diagnostician."

10 American Academy of Ophthalmology: Policy Statement: Learning Disabilities, Dyslexia, and Vision, 1981.

11 *Pediatrics* 2011; 127; e818.

12 *Symptomatic convergence insufficiency* is a diagnosis (metaphor) used to explain the inability to comfortably use the eyes at reading distance despite having the best glasses.

13 *Pediatrics* 2011; 127; e834.

14 Just as *neurotic* refers to symptoms not predicted by a currently accepted explanation by medicine, *placebo* refers to any cure not predicted by the same caliber of explanation. In the convergence insufficiency study, the placebo group was given sham vision therapy to be sure patients weren't committing the ultimate crime of getting better for the wrong reason.

15 (Coyne 2009), (Dawkins, The Greatest Show on Earth: The Evidence for Evolution 2009), and (Miller 2008).

16 (Abrams 1968), p. 1268. Quote from Thomas Hobbs' *Leviathan*.

17 (Rummel 2008), p. 1.

18 Both Marxism and Social Darwinism were initially touted as *scientific*.

19 In a Feb. 15, 1676, letter to fellow scientist Robert Hooke, *http//en.wikiquote. org/wiki/Isaac-Newton*.

1. Skepticism without a Doubt

20 Hume 1980, p. 19.

21 [22] For a complete discussion on philosophical or radical skepticism see Unger 1975.

22 [23] Shermer, *The Believing Brain, From Ghosts and Gods to Politics and Conspiracies—How we Construct Beliefs and Reinforce Them as Truths* 2011, pp. 2-3.

23 [24] Shermer, *Skepticism 101: How to Think Like a Scientist, Course Guidebook* 2013, p. 4.

24 (Aristotle 1984), p. 2154. From *Rhetoric, Book I*.

25 Huxley, Thomas, "Mr. Darwin's Critics" (1871), in *Collected Essays (1893-4), Volume II*, p. 165.

26 Samuel Butler's *Notebooks* (1951, p.233) quoted in (Andrews 1993), p. 807.

27 (R. P. Feynman 1999), p. 187.

28 Quoted in *Observer* (London, Aug. 1, 1982), quoted in (Andrews 1993), p. 809.

29 J. B. Birks, *Rutherford a Manchester*, London, 1962, in *The Oxford Dictionary of Quotations*, 1996, p. 108.

30 Francesco Petrarch, *On His Own Ignorance and That of Many Others*, transl. Hans Nachod, in *The Renaissance Philosophy of Man*, quoted in "The cultural authority of natural history in early modern Europe" (Alexander 2010), p. 11.

31 Hyperbole—other natural philosophers had written on history before Darwin, but in Gould's words, "The *Origin* focuses upon the establishment of a methodology for making inferences about history from features of modern organisms" (Gould, The Structure of Evolutionary Theory 2002), p. 103.

32 Functional magnetic resonance imaging (fMRI) is a procedure that studies changes in the blood flow of the brain to divine how we think. The procedure uses statistics to generate a colorful picture of brain activity.

33 Again, a complete recipe for mixing profits and science is provided in Ben Goldacre's delightful *Bad Pharma: How Drug Companies Mislead Doctors and Harm Patients* (Goldacre 2013).

34 (Nagel, The View From Nowhere 1989), p. 9.

35 (Andrews 1993): 809.

36 For a review, you'll find *Biology and Ideology: From Descartes to Darwin* enlightening (Alexander 2010).

2. Exploring Biomythology

37 (Lahave 1999), pp. 117-121.

38 Karl Marx (1818-1883) was a philosopher who told a passionate story in which some property was more equal than others, intellectual property remaining private, industrial capital deserving redistribution. Marx was passionate about the Robin Hood story, but he was careful not to carry the sole manuscript for *Das Kapital* through Sherwood Forest. He was not about to share his byline with those whose creative abilities fell short of his own. He saved alienation for factory workers, not readers of philosophy.

39 Ibid., p. 118.

40 (Dawkins, The God Delusion 2006), p. 135.

41 (McGinn, The Mysterious Flame: Conscious Minds in a Material World 1999), p. xi.

42 (Noë, Action in Perception 2004), p. 231.

43 Madrigal, Alexis, "Scanning Dead Salmon in fMRI Machine Highlights Risk of Red Herrings," *Wired*, Sept. 18, 2009. *http://www.wired.com/2009/09/fmrsalmon/*

44 In truth, this statement was either ironic or not.

45 According to Gould, Teilhard may have created a hoax that influenced scientists in the wrong direction for half a century. See "The Piltdown Conspiracy," in (Gould, The Richness of Life: The Essential Stephen Jay Gould, Edited by Steven Rose 2007), pp. 182-204.

46 Quoted in (Koch 2012), p. 133.

47 (Alexander 2010), p. 22.

48 (The Shorter Oxford English Dictionary 2007).

49 "Isaac Newton's Theory of the Universe," (Goldman 2006), Lecture 4, p. 58.

50 In Carter Kaplan, *Critical Synoptics: Menippean Satire and the Analysis of Intellectual Mythology* (New Jersey Associated University Presses, 200), p. 29, quoted in "Wittgenstein: scientific mythology," Ex Machina. *http://exmachina-tmr.tumblr.com/post/25833713606/wittgenstein-scientific-mythology*

51 (Kuhn 1996), p. 2.

52 "The Philosophy of Science" in C. A. Mace (ed.), *British Philosophy in the Mid-Century*, in Angela Partington (ed.), *The Oxford Dictionary of Quotations*, Oxford Univerity Press, 1996.

53 (Feyerabend 2010), p. 14.

54 (Lewis 1999), p. 13.

55 Fourier's work is described in the lecture "Science Comes of Age in the 19th Century," (Goldman 2006).

56 (Shermer, Why Darwin Matters, The Case Against Intelligent Design 2006), p. 40.

57 Miller, *Only a Theory: Evolution and the Battle for America's Soul,* 2008, p. 215.

3. Schools, Courts, and Imbeciles

58 (Dennett, Breaking the Spell: Religion as a Natural Phenomenon 2006), p. 327.

59 [62] Skepticism should also be applied to my interpretation of the great philosophers.

60 "Idealism," (Blackburn 2008), p. 177.

61 (Greenblatt 2012), p. 5.

62 Ibid.

63 Ibid., p. 6.

64 5.61, (Wittgenstein, Tractatus Logico-Philosophicus 1918).

65 (Bacon 2000), p. 19.

66 (Bacon 2000), p. 37.

67 Quoted in MacIntosh, J.J. and Anstey, Peter, "Robert Boyle," The Stanford Encyclopedia of Philosophy, Fall 2010 Edition, Edward N. Zalta (ed.), *http://plato.stanford.edu/achives/fall2010/entries/boyle,* p. 4.

68 Quoted in (Alexander 2010), p. 19.

69 Peter Harrison, "The cultural authority of natural history in early modern Europe," (Alexander 2010), p. 27.

70 (Gregory 2008), p. 110.

71 Google Digital Copy, (Newton 1756), p. 1.

72 Ibid., p. 5.

73 (Nagel, Mind and Cosmos: Why the Materialist Neo-Darwinian Conception of Nature is Almost Certainly False 2012), p. 128.

74 Ibid., p. 123.

75 (Kuhn 1996), p. 172.

76 (Descartes 2001), April 2, 2014, *www.bartleby.com/34*

77 (Bacon 2000), p. 61.

78 Quoted in (Solomon, Spirituality for the Skeptic: The Thoughtful Love of Life 2007), p. 65.

79 Chapter 8, (Rosenblum, Bruce; Kuttner, Fred 2008).

80 (Partington, Angela, ed 1996), p. 358.

81 The principle source of my Darwin quotes is (Darwin, On The Origin of Species: The Illustrated Edition, David Quammen, General Editor 2008).

82 (Miller 2008), p. 2.

83 (Darwin, On The Origin of Species: The Illustrated Edition, David Quammen, General Editor 2008), p. 513.

84 "Amicus Curiae Brief of 72 Nobel Laureates, 17 State Academies of Science, and 7 Other Scientific Organizations, in Support of Appellees, Robert A Klayman, Walter B. Slocombe, Jeffrey S. Lehman, Beth Shapiro Kaufman, Caplin and Drysdale, Charted One Thomas Circle, N.S., Washington, D.

C. 20005, (202) 862-5000, Attorneys for Amici Curiae," *http://www.talkorigins.org/faqs/edwards-v-aguillard/amicus1.html*.

85 (R. P. Feynman 1999), p. 187.

86 (Coyne 2009), p. xvii.

87 (Arbesman 2012), p. 2.

88 (Ruse and Travis 2009), p. ix.

89 (Shermer, Why Darwin Matters, The Case Against Intelligent Design 2006), p. xxii.

90 (Ruse and Travis 2009), p. vii.

91 (Miller 2008), p. 3.

92 (Darwin, Darwin: The Indelible Stamp: The Evolution of an Idea. Edited, with commentary by James D. Watson 2005), p. vii.

93 Reported in "Fury at DNA pioneer's theory: Africans are less intelligent than Westerners," *http://www.independent.co.uk/news/science/fury-at-dna-pioneers-theory-africans-are-less-intelligent-than-westerners-394898.html*.

94 (Coyne 2009), p. xiii.

95 (Charlesworth 2003), p. 2.

96 See Essay 14: "Rock of Evidence."

97 (Feyerabend 2010), p. 126.

98 See Chapter 1, "Mary Dyer Must Die," (Willis 2007).

99 Nature 484, pp. 304-306, 2012.

100 Sorry. Couldn't resist. Nothing personal.

4. Intelligent Designs

101 (Darwin, On The Origin of Species: The Illustrated Edition, David Quammen, General Editor 2008), p. 98.

102 Ibid., p. 99.

103 The ideas for this whole comparison of Darwin and Smith, quotes and all, are presented in (Gould, The Structure of Evolutionary Theory 2002), pp. 123-124.

104 (Gregory 2008), p. 382.

105 F. Darwin, 1887, quoted in (Gould, The Structure of Evolutionary Theory 2002), p. 116.

106 Electronic Scholarly Publishing Project, 1998, *http://www.esp.org*, (Malthus 1798)

107 (Brooks 1984), p. 54.

108 Ibid., p. 102.

109 (van WYHE 2009)

110 (Darwin, On The Origin of Species: The Illustrated Edition, David Quammen, General Editor 2008), p. v.

111 (P. Johnson 2012), p. 81.

112 (Darwin, On The Origin of Species: The Illustrated Edition, David Quammen, General Editor 2008), p. x.

113 (Feyerabend 2010), p. 204.

David Cook

114 (Gregory 2008), p. 490.
115 *http://www.gutenberg.org/files/25308/25308-h/25308-h.htm*, p. 3.
116 (Darwin, Darwin: The Indelible Stamp: The Evolution of an Idea. Edited, with commentary by James D. Watson 2005), p. vi.

5. Evolutionary Equivocation
117 (Cave 2012), p. 281.
118 (Dawkins, The God Delusion 2006), p. 32.
119 May 15, 2013, *http://www.nature.com/news/science-in-schools-1.12979?WT.
ec_id=NATURE-20130516*
120 (Dawkins, The Greatest Show on Earth: The Evidence for Evolution 2009), p. 429.
121 (Lisle 2010), Chapter 2.
122 (Coyne 2009), p. xvi.
123 Ibid., p. xvii.
124 (Pinker 2009), p. 155.
125 Ibid., p. 157.
126 Ibid., p. 162.
127 (Pinker 2009), p. 162.
128 Quoted in (P. E. Johnson 1993), p. 20.
129 Sober, *Philosophy of Biology,* 2000, p. 70.
130 Ibid., p. 74.
131 (Gould, The Structure of Evolutionary Theory 2002), p. 369.
132 (Gould, The Structure of Evolutionary Theory 2002), p. 409.
133 Ibid., p. 406.
134 From Batson, W., *Materials for the Study of Variation,* 1894, London, Macmillian, quoted in (Gould, The Structure of Evolutionary Theory 2002), p. 407.
135 (Pinker 2009), p. 427.
136 Ibid., p. 165.
137 (Fodor, J. and Piattelli-Palmarini, M. 2010), p. xiv.
138 Also called "internalists."
139 (Dawkins, The God Delusion 2006), p. 293.
140 Ibid., p. 294.
141 (Gould, The Structure of Evolutionary Theory 2002), p. 452.
142 Ibid., p. 454.
143 (Darwin, On The Origin of Species: The Illustrated Edition, David Quammen, General Editor 2008), p. 99.
144 Ibid., p. 196.
145 Ibid., p. 191.
146 (Coyne 2009), p. 223.
147 (Fodor, J. and Piattelli-Palmarini, M. 2010), p. 54.
148 Whitfield, John, "Postmodern evolution?" *Nature* Vol. 455, 18 Sept. 2008, pp. 281-284.

149 Darwin is seen as the father of natural selection. Mendel, who performed famous experiments on plants, is seen as the father of genetics. The cocktail of natural selection mixed with genetics is known as "the Modern Synthesis."

150 Ibid., p. 284.

6. Murder by Ugly Fact

151 (Dawkins, The Greatest Show on Earth: The Evidence for Evolution 2009), p. 100.

152 (Coyne 2009), p. 223.

153 (Gopnick 2009), p. 185.

154 An *anomaly* is a disappointment to a theory. Whenever a "fact" breaks down, rather than jettison the fact, we call the exception an *anomaly*. An *anomaly* is the beauty cream used to cover the warts of ugly facts. It is a fact that children are born with ten toes. It is an anomaly when they are born with twelve.

155 (Shermer, Why Darwin Matters, The Case Against Intelligent Design 2006), p. 16.

156 (Dawkins, The Greatest Show on Earth: The Evidence for Evolution 2009), p. 147.

157 (Darwin, On The Origin of Species: The Illustrated Edition, David Quammen, General Editor 2008), p. 336.

158 (Popper 2002), p. 63.

159 (Nagel, Mind and Cosmos: Why the Materialist Neo-Darwinian Conception of Nature is Almost Certainly False 2012), p. 12.

160 (Pinker 2009), p. 171.

161 (Fodor, J. and Piattelli-Palmarini, M. 2010), p. 45.

162 *Ad hoc*: made up or improvised for the special purpose at hand.

163 (Popper 2002), pp. 60-61.

164 (Miller 2008), p. 181.

165 (J. Hall 2005), p. 142.

166 (Dawkins, The God Delusion 2006), p. 127.

167 (Bacon 2000), p. 43.

168 (Darwin, On The Origin of Species: The Illustrated Edition, David Quammen, General Editor 2008), pp. 502-503.

169 The rainbow goes from red to orange to yellow, etc., all the way to violet. Just beyond the visible violet is the *ultraviolet*, with wavelengths too short to be perceived by humans.

170 (Rosenblum, Bruce; Kuttner, Fred 2008), p. 55.

171 Quoted in (Kuhn 1996), p. 151.

172 (Kuhn 1996), pp. 17-18.

173 (Kuhn 1996), p. 147. It should be noted that Kuhn is not stating as a hard-and-fast rule that you must have a new theory before you are allowed to find fault with an old one that is currently closing the door on original thinking. He is merely illustrating how paradigm shifts normally work. The

best way to break paradigm is like the best way to break a habit—with a new one.
174 (Aristotle 1984), p. 1588.

7. Is Darwinism a Religion?

175 See Chapter 5, "Murder by Ugly Fact."
176 (Gopnick 2009), p. 185.
177 (Harris, Letter to a Christian Nation 2006), p. 18.
178 (Gould, The Structure of Evolutionary Theory 2002), p. 148.
179 (Darwin, On The Origin of Species: The Illustrated Edition, David Quammen, General Editor 2008), p. 8.
180 (Jablonka, Eva and Lamb, Marion J. 2006), p. 9.
181 Sir Charles Lyell (1797-1875): a Scottish geologist and one of the most respected scientists of Darwin's age. Lyle's most famous book, *Principles of Geology*, came out in 1830 and was studied by Darwin during the five-year-long voyage of the *Beagle*. Lyell argued for gradual changes in the geology of the earth. From Lyell, Darwin borrowed the gradualism that would define his view of natural selection.
182 Georges Cuvier (1769-1832): French naturalist and geologist who made a name for himself after the French Revolution.
183 Jean-Baptiste Lamarck (1744-1829): French evolutionist who theorized that changes an organism made to adapt to the environment could be passed on to the next generation.
184 While genetics is largely about changes in the sequence of genes, *epigenetics* is the theory that genes may be turned "on" (expressed) by the environment without changing their sequence. Once the "switch" has been thrown, the change in gene expression may be passed on to the next generation. For instance, once cigarettes throw the cancer switch, the next generation may be prone to cancer. If epigenetics were true, then the Darwinians who ridiculed Lamarck throughout the twentieth century may themselves be ridiculed throughout the twenty-first century. Such is the price of confusing explanation with truth.
185 (Dawkins, The Greatest Show on Earth: The Evidence for Evolution 2009), p. 18.
186 (Dawkins, The God Delusion 2006), p. 283.
187 As we will discuss in the chapter "Rock of Evidence."
188 Aristotle, Book II, Chapter 8. All these thoughts on Aristotle were inspired by David Roochnik's lecture "Aristotle's Four Causes" in his Great Courses' *An Introduction to Greek Philosophy*.
189 (Kuhn 1996), p. 171-172.
190 Teleology can, but does not have to, be based on theism or vitalism. Whether or not natural selection is teleological is a "hotly contested question." For one treatment of the subject see Lennox, James, "Darwinism", The *Stanford Encyclopedia of Philosophy*, Fall 2010 Edition, Edward N. Zalta (ed.), *http://plato.stanford.edu/archives/fall2010/entries/Darwinism*.

191 Evo-devo: evolutionary developmental biology, a discipline that argues that the arrival of a species is a direct result of genetic changes in the process of development, rather than competition, as Darwin had believed.

192 (Gould, The Structure of Evolutionary Theory 2002), p. 352.

193 (Nagel, Mind and Cosmos: Why the Materialist Neo-Darwinian Conception of Nature is Almost Certainly False 2012), p. 123.

194 *Absurd* may be defined as out of harmony with reason. If eternity falls outside of time, it would not be reasonable for eternity to fall *inside* of time. According to Kierkegaard, "The absurd is that the eternal truth has entered time, that God has entered existence." Kierkegaard believed that to accept this mismatch, this outside being inside, required faith.

195 Kierkegaard, Søren, *Concluding Unscientific Postscript*, quoted in (Landesman, Charles and Meeks, Roblin 2003), p. 269.

196 Ibid., p. 265.

197 Again, *paradigm* is Kuhn's term for a fundamental theory during a period of history. For Newton, space and time were absolute; for Einstein, they were not. Each way of looking at space was a fundamental theory, a paradigm.

198 (Shermer, Why Darwin Matters, The Case Against Intelligent Design 2006), p. 161.

199 (Sagan 1997), p. 29.

200 Ibid., p. 27.

201 Quoted in (Dawkins, The God Delusion 2006), p. 12.

202 Ibid.

203 (Miller 2008), p. 221.

204 (Smith 1995), p. 893.

205 (Damasio 2005), p. xii.

206 Or any other false alternative you wish.

207 (Landesman, Charles and Meeks, Roblin 2003), p. 260.

208 (Tillich 2009), pp. 1-2.

209 (Dennett, Intuition Pumps and Other Tools for Thinking 2013), p. 21.

210 (Dawkins, The God Delusion 2006), p. 117.

211 (Pinker 2009), p. 155.

212 (Ruse and Travis 2009), Forward, Edward O. Wilson, p. vii.

213 (Miller 2008), p. 134.

214 (Coyne 2009), p. 233.

215 (Shermer, Why Darwin Matters, The Case Against Intelligent Design 2006), p. 161.

216 (Dennett, Darwin's Dangerous Idea: Evolution and the Meaning of Life 1995), p. 21.

8. The Rhetoric of Objectivity

217 (Aristotle 1984), p. 2,155.

218 (Ryle 2002), p. 15.

219 (Koch 2012), p. 8.

220 Thomas Sprat, from "The History of the Royal Society," *http://www.wnorton. com/college/english/nael/noa/pdf/27636_17th_U38_Sprat-1-6.pdf.*

221 (Darwin, On The Origin of Species: The Illustrated Edition, David Quammen, General Editor 2008), p. 513.

222 (Fodor, J. and Piattelli-Palmarini, M. 2010), p. 152.

223 Another argument of structuralism: there are internal factors which guide change over time.

224 See the essay "Musical Definitions."

225 *Tu quoque* fallacy—enjoy.

226 (Darwin, On The Origin of Species: The Illustrated Edition, David Quammen, General Editor 2008), p. 77.

227 From Bacon's "Novum Organum, The New Organon," which takes its title from Aristotle's work on logic, the "Organon" or "Instrument for Rational Thinking" (Bacon 2000).

228 Gould quotes from the sixth edition of *Origin*, 1872, p. 395. (Gould, The Structure of Evolutionary Theory 2002), p. 147.

229 (Harris, The Moral Landscape: How Science Can Determine Human Values 2010), p. 47.

230 (Coyne 2009), p. 16.

231 (Darwin, On The Origin of Species: The Illustrated Edition, David Quammen, General Editor 2008), p. 174.

232 Ibid., pp. 189-191.

233 Ibid., p. 174.

234 Ibid., p. 181.

235 (Dennett, Intuition Pumps and Other Tools for Thinking 2013), p. 25.

236 Ibid., p. 26.

237 Ibid., p. 51.

238 Chapter 3: Intelligent Designs

239 (Wittgenstein, Philosophical Investigations, The German text, with an English translation by G.E.M. Anscombe, Revised 4th edition by P.M.S. Hacker and Joachim Schulte 2009), sec. 350.

9. Musical Definitions

240 I am a pseudo-pacifist, meaning I deplore killing but am more than happy to enjoy the fruits of a country created by war. A real pacifist would give his property back to the Native Americans.

241 (Rummel 2008), p. 1.

242 (Wittgenstein 2009), p. 85.

243 (Nagasawa 2008), p. 23.

244 Ibid., p. 24.

245 Popular aphorism quoted in (Blackmore 2004), p. 219.

246 (Dawkins, The Selfish Gene 30th Anniversary Edition, 2006), p. 28.

247 (Shermer, Why Darwin Matters, The Case Against Intelligent Design 2006), p. 6.

248 (Dawkins, The Greatest Show on Earth: The Evidence for Evolution 2009), p. 303.

249 I am, of course, using my own form of biomythology, transporting *scientific* concepts out of context to blunt common sense.

250 (The Oxford American College Dictionary, 2002).

251 R. Dawkins, *The Blind Watchmaker*, page 140, quoted in (A. Plantinga 2011), p. 26.

252 (Shermer, Why Darwin Matters, The Case Against Intelligent Design 2006), p. 123.

253 (Shermer, The Believing Brain, From Ghosts and Gods to Politics and Conspiracies--How we Construct Beliefs and Reinforce Them as Truths 2011), p. 177.

254 See Essay 183, "The Madness of Shakespeare's Method."

10. Rock of Evidence

255 These tests are detailed in the lecture "Minds Possessed—Witchery and the Search for Explanation," (Robison 1997).

256 (Encarta World English Dictionary 1999).

257 Diogenes Laertius, from *Pyrrho*, in (Landesman, Charles and Meeks, Roblin 2003), pp. 23-29.

258 (Dawkins, The Greatest Show on Earth: The Evidence for Evolution 2009), p. 15.

259 (Bacon 2000), pp. 18-19.

260 Ibid.

261 Ibid.

262 (Goldman 2006), p. 44.

263 (Feyerabend 2010), p. 50.

264 Ibid., p. 51.

265 Ibid., p. 127.

266 Ibid., p. 132.

267 Ibid., p. 134.

268 See (Hume 1969), pp. 520-521.

269 Ibid., p. 85.

270 From a letter to Father Castelli quoted on p. 2 in College of Optometrists, Galileo Galilei (1564-1642). May 28, 2013, *http://www.college-optometrists. org/en/knowledge-centre/museyeum/online/_exhibitions/o...*

271 (Bacon 2000), p. 18.

272 Ibid.

273 John P.A. Ioannidis, *Plos Medicine, www.plosmedicine.org*, August 2005, Volume 2, Issue 8, e124.

274 The following stories about Boyle were taken from pp. 151-153 of (Gregory 2008).

275 (Gregory 2008), p. 153.

276 Ibid.

277 (Jablonka, Eva and Lamb, Marion J. 2006), p. 25.

278 Ibid., p. 58.
279 Ibid., p. 73.
280 (Dawkins, The Greatest Show on Earth: The Evidence for Evolution 2009), pp. 116-131.
281 (Noë, Action in Perception 2004), p. 1.
282 Hubel, D. H., and Wiesel, T. N., "Receptive fields, binocular interaction and functional architecture in the cat's visual cortex," J. Physiol, 1962; 160: pp. 106-154.
283 Pediatric Eye Disease Group, "Randomized trial of treatment of amblyopia in children ages 7 to 17 years," Archives of Ophthalmol. 2005; 123(4): 437-447.
284 Freedman LP, Cockburn IM, Simcoe TS (2015) "The Economics of Reproducibility in Preclinical Research." PLoS Biol 13 (6): e1002165. DOI: *10.1371/journal.pbio.1002165*
285 Chapter 17: Free Will Versus Free Excuse
286 Ferguson, Andrew, "Making It All Up, The behavioral sciences scandal," The Weekly Standard, October 19, 2015, 21(6). *http://www.weeklystandard.com/articles/making-it-all_1042807.html*
287 (Hazen 2001), p. 2.
288 (P. Johnson 2012), p. 67.
289 (Darwin, On The Origin of Species: The Illustrated Edition, David Quammen, General Editor 2008), p. 190.
290 (Coyne 2009), p. 142.

11. Fallacy to a Higher Power

291 (Feyerabend), p. 9.
292 Book II, Part III, Section III, (Hume, A Treatise of Human Nature, Edited by Ernest C Mossner 1969), p. 462.
293 See (Grim 2013), pp. 111-127.
294 It wasn't. I created the platitude for rhetorical purposes.
295 (Lisle 2010), p. 65.
296 (Goldman 2006), p. 34.
297 (Goldman 2006). Ibid., p. 35.
298 Snyder, Laura J., William Whewell, The Stanford Encyclopedia of Philosophy (Winter 2012 Edition) Edward N. Zalta (ed.), *http://plato.stanford.edu./archives/win2012/entries/whewell.*

12. The Road to Truth

299 Republished in (Landesman, Charles and Meeks, Roblin 2003), p. 347.
300 (Harris, Letter to a Christian Nation 2006), p. 64.
301 (Shermer, The Believing Brain, From Ghosts and Gods to Politics and Conspiracies--How we Construct Beliefs and Reinforce Them as Truths 2011), p. 7.
302 (Coyne 2009), p. xvi.
303 Ibid., p. 16.

304 (Koch 2012), p. 27.

305 Ibid.

306 (Dawkins, The Greatest Show on Earth: The Evidence for Evolution 2009), p. 17.

307 (Nagel, Mind and Cosmos: Why the Materialist Neo-Darwinian Conception of Nature is Almost Certainly False 2012), p. 3.

308 Section X, Part I, (Hume, An Enquiry Concerning Human Understanding: A letter from a Gentleman to his Friend in Edinburgh, Edited by Eric Steinberg 1980), p. 77.

309 (Goldman 2006), p. 14.

310 (Gregory 2008), p. 87.

311 (Cox 2010), pp. xi-xii.

312 Essay 2, "Schools, Courts, and Imbeciles."

313 (Dawkins, The Greatest Show on Earth: The Evidence for Evolution 2009), p. 13.

314 (Feynman, Leighton and Sands 2010), Chapter 4, Section 1.

315 (Bacon 2000), p. 36.

316 Ibid., p. 16.

317 (Hume, An Enquiry Concerning Human Understanding: A letter from a Gentleman to his Friend in Edinburgh, Edited by Eric Steinberg 1980), p. 24.

318 See Introduction and Chapter 1 for the definitions.

319 Lars Muckli, Marcus J. Naumer, Wolf Singer, "Bilateral visual field maps in a patient with only one hemisphere," *www.pnas.org/cgi/doi/10.1073/pnas.0809688106*

320 (Charlesworth 2003), p. 1.

321 (Gould, The Structure of Evolutionary Theory 2002), p. 1.

322 (Krukonis and Barr 2008), p. 3.

323 Essay 15: Rhetorical Numbers and Apes.

324 (Dawkins, The Greatest Show on Earth: The Evidence for Evolution 2009), p. 9.

325 (Coyne 2009), p. xvii.

326 For a 500-page review of the subject, see (Gould, The Structure of Evolutionary Theory 2002).

327 (Fodor, J. and Piattelli-Palmarini, M. 2010), p. xx.

328 (Nagel, Mind and Cosmos: Why the Materialist Neo-Darwinian Conception of Nature is Almost Certainly False 2012), p. 128.

329 (Popper 2002), p. 5.

330 Ibid., p. 4.

331 Defining love, of course, is not so easy. Is it love to place others before yourself? Is it love to be willing to die (but not kill) for the salvation of others? Is it love to kiss a girl who science assures us is part of a colony that is 90 percent bacteria?

332 Again, excuse me: my genome made me do it.

13. Some Falsehoods are More Equal than Others

333 (Dawkins, The God Delusion 2006), p. 52.

334 Yes, I am equivocating.

335 Ibid., p. 348.

336 Ibid., p. 349.

337 (Gould, The Structure of Evolutionary Theory 2002), p. 407.

338 (Dawkins, The God Delusion 2006), p. 130.

339 Asimov, Isaac, "The Relativity of Wrong," The Skeptical Inquirer, Fall 1989, Vol. 14, No. 1, pp. 35-44. *http://chem.tufts.edu/answersinscience/relativityofwrong.htm*

14. The Madness in Shakespeare's Method

340 These reflections on Shakespeare and behavioralism, I owe to (McGinn, Shakespeare's Philosophy: Discovering the Meaning Behind the Plays 2007).

341 (Bacon 2000), p. 8.

342 (Dawkins, The Magic of Reality: How We Know What's Really True 2012), pp. 21-22.

343 (Feyerabend 2010), p. 7.

344 (Bacon 2000), p. 22.

345 (Israel 1998), p. 465.

15. Rhetorical Numbers and Apes

346 We will revisit fideism in our final chapter.

347 (R. P. Feynman 1999), p. 187.

348 Quoted in (Andrews 1993), p. 870.

349 Ibid.

350 (Gould, The Structure of Evolutionary Theory 2002), p. 69.

351 (Dennett, Intuition Pumps and Other Tools for Thinking 2013), p. 100.

352 Of course not. They didn't see the light from the same perspective at the same time—Aristotle covered his bases, and biases.

353 (Silvertown 2008), p. 59.

354 Ibid., pp. 105-106.

355 Ibid., p. 107.

356 Ibid., p. 15.

357 (Gould, The Structure of Evolutionary Theory 2002), p. 512.

358 Shakespeare's priority was brought to my mind by (McGinn, Shakespeare's Philosophy: Discovering the Meaning Behind the Plays 2007), p. 149.

359 Just kidding.

360 (Dennett, Intuition Pumps and Other Tools for Thinking 2013), p. 37.

361 Quoted in (A. Plantinga 2011), page 316.

16. Art, Nature, and Reconstruction

362 (Ash 2005), dust jacket flap.

363 A legend I just created for rhetorical purposes.

364 As explained in (Gregory 2008), p. 360.

365 Quoted, page 361, (Gregory 2008), p. 361.
366 Quoted, page 481, (Gould, The Structure of Evolutionary Theory 2002), p. 481.
367 Stokes, W.L. Essential of Earth History, 1973, Englewood Cliffs, NJ, Prentice-Hall, p. 37. Quoted, page 483, (Gould, The Structure of Evolutionary Theory 2002), p. 483.
368 (Woodward 1948), *http://www.clarku.edu/~ Piltdown/map_report_finds/ earliest_english.html*, p. ix.
369 (Johanson 2010), p. 3.
370 (Woodward 1948), p. 55.
371 Ibid., pp. 63-64.
372 Ibid., p. 65.
373 Ibid., p. 27.
374 Ibid., p. 56.
375 Ibid., p. 88.
376 Ibid., p. 74.
377 Ibid., p. xii.

17. The Ghost in the Paradigm
378 (Kuhn 1996), p. 10.
379 Ibid., p. 6.
380 Ibid., p. 8.
381 (Dawkins, The Greatest Show on Earth: The Evidence for Evolution 2009), p. 17.
382 "Wallace's Letters Online," Natural History Museum. *http//www.nhm.ac.uk/ research-curation/scientific-resources/collections/library-collection/ wallace-letters-online/373/373/T/details.html#3.*
383 (Kuhn 1996), p. 173.
384 (Eccles 1994), p. 7.
385 (J. D. Watson 2001), p. 7.
386 (Crick 1995), p. 3.
387 Ibid.
388 (Shermer, The Believing Brain, From Ghosts and Gods to Politics and Conspiracies--How we Construct Beliefs and Reinforce Them as Truths 2011), p. 156.
389 "The Monkey Business Illusion," *http://www.youtube.com/ watch?v+IGQmdoK_ZfY.*
390 This metaphor isn't close enough for philosophy. Applying "Musical Definitions" (see Chapter 9), it could be argued that conscious mental states are not the mental states to be considered, that the great unconscious—much as Darwin (see Chapter 8, "Murder by Ugly Fact") used the "imperfect geological record" to save the identity theory. Add this to the imperfect neural record and we can, with no risk, always posit that some function is equal to some function. Since neither the great unconscious nor the imperfection of our understanding of the neurological

record can be falsified, the theory is not science. Similarly, the skeptic should be aware that my own metaphor qualifies (as does most of this book) as natural rhetoric designed to trivialize the contributions of any number of dedicated thinkers—an accolade that will never be applied to me.

391 (M. Robinson 2010), pp. 49-50.

392 *Brain-Wise Studies in Neurophilosophy*, Cambridge, MA: MIT Press, 2002, p.1: quoted in (Noë, Out of Our Heads: Why You Are Not Your Brain, and Other Lessons from the Biology of Consciousness 2009), p. 6.

393 Ibid.

18. Free Will Versus Free Excuse

394 Those who argue that what is happening now is based entirely on what has happened in the past are determinists.

395 (Hume, An Enquiry Concerning Human Understanding: A letter from a Gentleman to his Friend in Edinburgh, Edited by Eric Steinberg 1980), p. 53.

396 Immanuel Kant, from *Critique of Practical Reason*, quoted in (Pereboom, Derk. ed. 1997), p. 103.

397 (Kane, A Contemporary Introduction ot Free Will 2005), p. 1.

398 Chisholm, Roderick, "Human Freedom and the Self," quoted in (Pereboom 1997), pp. 143-144.

399 (Harris, Free Will 2012), p. 13.

400 (Harris, Free Will 2012), p. 4.

401 (Harris, Letter to a Christian Nation 2006), p. 88.

402 See "The Method in Shakespeare's Madness."

403 For the details, see (Benjamin 2004).

404 See the last chapter's discussion on perceived quarter size and brain activity.

405 (Koch 2012), p. 78.

406 Ibid., p. 30.

407 Quoted in (Pereboom, Derk. ed. 1997), p. 57.

408 Quoted in (Pereboom, Derk. ed. 1997), p. 147.

409 Quoted in (Pereboom, Derk. ed. 1997), p. 152.

410 It is difficult to say what split-brain experiments really tell us about being human. The brain is damaged, eye movements are not allowed, the world is altered so that information can be presented to one visual field at a time. Thus, if perception is based on neurology, movement, and environment, altering all three might create the illusion of illusions and tell us little about intact persons moving in habitual environments.

411 (Pereboom 1997), p. 87.

19. The Golden Rule of Skepticism

412 Jesus's version of the rule is taken from the King James Bible, the other versions from (Bowker 1997), p. 88.

413 Shoving your enemy into the mud puddle, letting your enemy step in the puddle without warning it is eight feet deep—both violate the positive Golden Rule of Jesus and Mohammed. In its negative form, the rule forbids the shove, but not the silence.

414 (Rummel 2008).

415 (Wilson 2012), p. 187.

416 Ibid., p. 189.

417 (Harris, The Moral Landscape: How Science Can Determine Human Values 2010), pp. 1-2.

418 (D. N. Robinson 2007), p. 107.

419 (Hodges 2008), p. 60.

420 "Non-Overlapping Magisteria," (Gould, The Richness of Life: The Essential Stephen Jay Gould, Edited by Steven Rose 2007), p. 594.

421 (Gould, The Richness of Life: The Essential Stephen Jay Gould, Edited by Steven Rose 2007).

422 (Rummel 2008), pp. 1-2.

423 (Wittgenstein 2009) p. 109, p. 52.

424 See the "Seeing the Light" section in the "Rhetorical Numbers" chapter.

20. Stories of Skeptical Faith

425 Republished in (Landesman, Charles and Meeks, Roblin 2003), p. 102.

426 Education, Ch. 2 (1861), quoted in (Andrews 1993), p. 809.

427 Descartes, Rene, "Meditation VI," republished in (Landesman, Charles and Meeks, Roblin 2003), p. 283.

428 (Wittgenstein, Tractatus Logico-Philosophicus 1918), 6.371, p. 90.

430 Ain't no royalties in used books. Insist your friends buy a new copy.

Printed in the United States
By Bookmasters